The Indigenous Tribesmen of Neverland

a novel by

Michael Lyons

Vol. 2 of the Sextet
My Years of Apprenticeship at Love

HiT MoteL Press
www.hitmotel.com

Copyright 2010 by Michael Lyons
All rights reserved.
First Edition

Library of Congress Cataloging in publication Data

Lyons, Michael
    The Indigenous Tribesmen of Neverland
Vol.2 of the Sextet *My Years of Apprenticeship at Love*
I. Title

ISBN: 0-965584275
ISBN-13 978-065584272

Published by HiT MoteL Press

Designed by Michael Lyons

to
the Tribesmen

Table of Contents

The Indigenous Tribesmen of Neverland

Alive and Well in Austin 1
A Shuttlebus Named IF 10
Dogmen Down on the River 22
Like a Coyote in a Prairie Hail Storm 30
Fugue in A Minor 41
A Good Cry for Green Spring 53
Two On Pillows Dreaming 64
Home on the Range 70
Guarding Angels 83
Agent Felicity 89
The Night Auditor 99
Letter to a Campfire Girl 129
Tales of the Wild Seed Women 136
Some Correspondence with Dick Ache 170
A Room of One's Own 172
Austin Deep In the Wild Heart 183
In the Thirtieth Year of My Age 190
The Peter Pan Syndrome 193
Further Worry about the Situation 196
Good-bye Party 203
Heavy Metal Music at the Steel Works 209
Epilogue 214

Appendix: A Structural Analysis

Introduction 217
The Sorcerer's Apprentice 219
A Toast to Carlos Castaneda 235
The Theatre of the Tesseract (Intro) 238
The Subliminal Kid (a monologue) 241
The Theatre of the Tesseract (Notes) 262
Quarking the Cube 285
And we are way downstream in the light 313

> The clearest way
> to get into the Universe
> is to walk
> through a forest.
> --John Muir

The Indigenous Tribesmen of Neverland

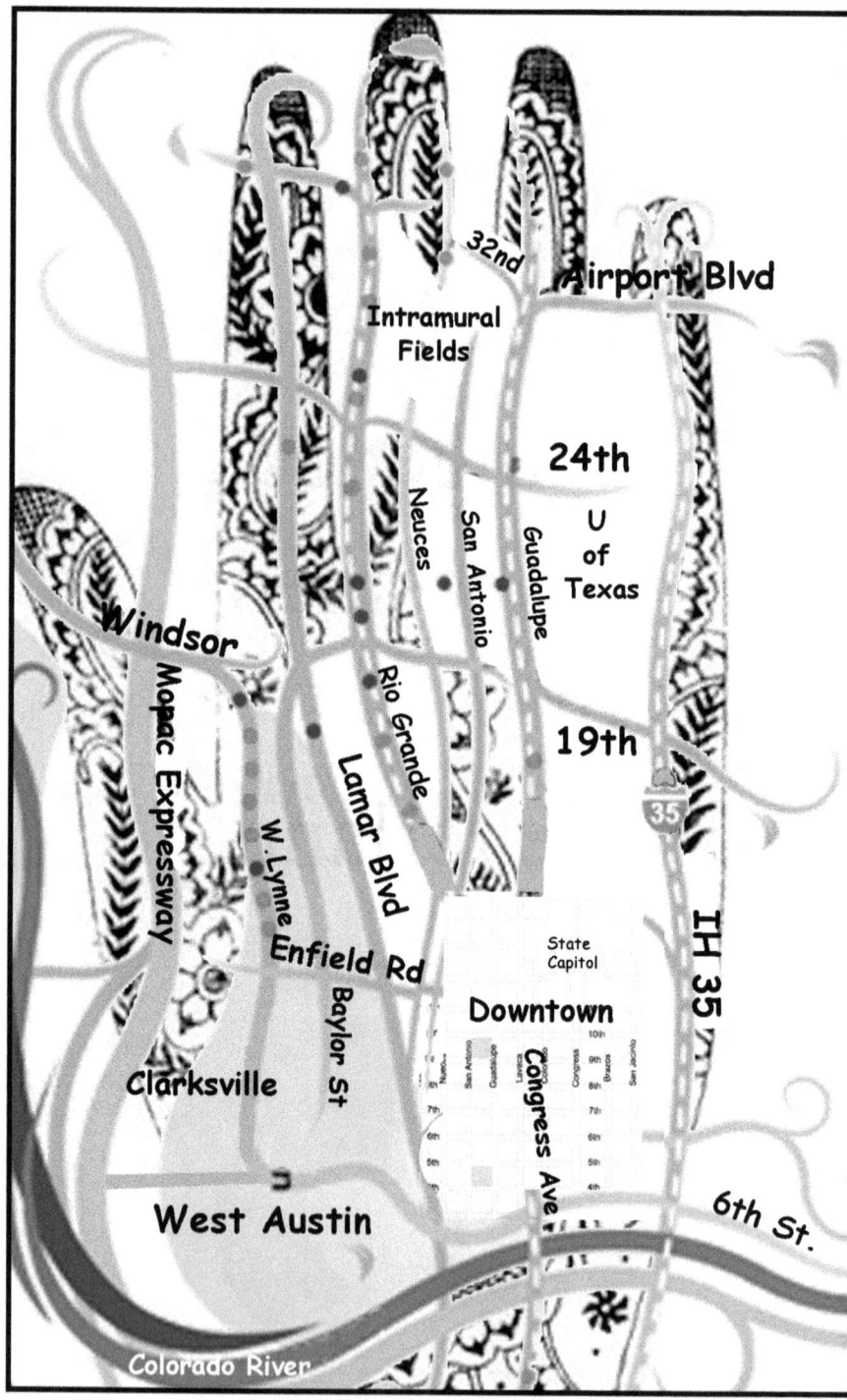

The Signed Hand House of Chance and Love

32nd
Airport Blvd
Intramural Fields
24th
Neuces
San Antonio
Guadalupe
U of Texas
Windsor
Rio Grande
19th
Mopac Expressway
Lamar Blvd
W. Lynne
35
IH 35
Enfield Rd
State Capitol
Downtown
Baylor St
Congress Ave
Clarksville
West Austin
6th St.
Colorado River

Alive and Well in Austin

It was raining down in Texas, falling down heavy that spring. It was the 30th year of my life and I was expecting some great mystery to unfold, but circumstances were anything but auspicious — they were tough. I was trying to get back on my feet from getting busted on the farm. I had once been a science teacher in college, but had been revisited by the unfulfilled restlessness of my youth and once again gone bumming around the country, hitchhiking and trying to be part of various utopian / creative / spiritual / healthy / artistic scenes. Now, compared to the material progress of the people I grew up with, and went to school with, I was slipping down the shadow path to a shoeless oblivion. Oh well, don't mean to be dramatic, it wasn't that bad. I had thought to get rich and have a freedom garden against the paraquat scare, and it grew into a huge marijuana farm operation and it got busted, and after I got out of jail, I had like 11 cents to my name and was homeless. I had to engage in various forms of couch-surfing and crashing at various friend's houses in a sort of indentured servitude labor exchange. Some kind of reckoning was required. My lawyer had to get paid, and the only job I could get was a tile setter's apprentice and I was grateful for that. Thus I had the money to put down the deposit and rent this great flat at the back of the big house at 1100 Baylor Street in Austin when it came available.

Our community of domiciles there was just off Lamar and within walking distance to down town. The layout of the big house and its associated little houses scattered around it and above it, occupied a hill, so that from any one of the houses you had a superb uninterrupted view of the skyline. Yet the

denizens therein were all screened by a forest of big old trees overhanging, so that you felt hidden and kind of invisible behind and under them. And in them too, for some of the trees had whole adult sized tree-houses, with verandas and swings in them. What's more, those tall (higher than an elephant's eye tall) reeds on the south side in front of Juliet Hapsburg-Bourbon's studio were so thick that no one but little pygmy ankle-biter children could possibly get through there. Yes, it was kind of wild and overgrown and shaggy, the way we liked it, all densely set about with trees and climbing vines and circuitous rock walks and stone steps with slab seats and grottoes set in great temple walls made of big white bricks quarried from the same place where the university got theirs, with little paleontological shell life forms in them. Talbot got all this masonry, left over from landscaping jobs or that he bought.

    It was one of the most interesting places of the many I had lived in Austin as a struggling student and bohemian slacker in that liberal college town, known as an oasis "Cambridge of the Plains" set in the capital of what could sometimes be a redneck hellhole state. While living there with a house mate named Wild Bill, I had a young girlfriend, Laura who was still in high school at the time.

    I was the last to rent there, I signed my name Walker Underwood to the lease for the big two story flat in the back of the big house. The flat was a run down fixer when I got in there and I didn't have a stick of furniture, just rolled out my sleeping bag on the floor in that big room up the stairs. The kitchen cabinets around the sink were pretty much rotted out from the leaks. I had a funny little blues song about it running through my head and would sing it sometimes, in that high falsetto voice of The Cream: "I need to find a washer, to stop up the leaky faucet of my doom. Gotta find me some kind of washer, to stop up the leaky faucet of my doo-oom."

The first thing I had to address was the big orange extension cord leading out of my kitchen window, across the central grounds to another house. I followed it snaking through some bush up to a neighbor's bungalow which was completely devoid of any paint, but with a lovely new porch overlooking the city. A sign tacked to a pillar read "Mane Event." Somewhere along the line they had appropriated their sign from a hair salon. The Mane Eventers were a bunch of hard drinking ne'er-do-wells. (It would be an insult to the concept to call them hippies — all of them had short hair, for one thing.) Little Billy and Big John lived there; I knew them from 33rd St. They were drinking buddies. Billy was really smart and funny; we used to talk about calculus, but he was always getting beaten up or roughed up by John. I had to pick him up off the kitchen floor there more than once. Sarah was Billy's old lady and she had a little stone house out back. Apparently the power company had run out of patience with each of them as none of them could get the electricity turned on.

They were fun to drink a beer with now and then — we were cool, even though I too had to give them notice to get their own electricity. This did push them to somehow get re-electrified.

And later I brought in Wild Bill with all his dogs after his wife told him he had to go, so she could get a paying roommate.

I was glad to have a roof over my head, especially with all the rains coming down in Texas in the spring time. No lie, the rains that year were particularly horrendous. We had every kind of rain: there was a street swallowing rain, a gully washer rain, there was a creek cresting rain, a river swelling rain, a deluge; some days it was like Niagara Falls falling on our heads. There was more than enough rain to feed the plants.

And the thunder and the lightning! I liked that. My father liked a good storm and I did too and here on the edge of tornado alley was a good place to see them. There would be lightning strikes on the buildings down town, and on the castle at the top our hill. We called it a castle, because it had towers with crenellated turrets like battlements, and was all made out of white limestone, and it had a big cement veranda with fluted stone columns. The castle held the hill top surrounded by a big grounds of green grass inside a white brick and wrought iron fence.

You'd see the lightning short-circuiting the great big dark tumultuous clouds over the city with their fluxing capacitance welling in the somber ominous sky — lightning illuminating the dark oak trees outside my window in the night — then hear the angry ripping crack of window-rattling thunder close by, like it was some kind of dijnn running around outside the big lonely house chasing demons in the dark and it might just come in GRAB you too.

It was a bit scary with all that rain; sometimes it would rain for a whole week. I'd be cooped up inside looking out the window at the city in the rain falling on the windowpane. That rain could be ferocious, spraying like a fire hose; falling like a river from the sky. It got downright biblical: God told Noah, 'Build me an ark. For the covenant has been broken and I am going to wash the filth from the streets of the earth.'

The water is rising. The lake is swallowing trees. Who would have thought that there could be so much water! Water everywhere — now — but come summer it would get so dry they sometimes sprayed the roads to keep the dust down. Then the next day the sun would be out. It was a wee bit lonely up there in that big flat, and I was glad to have Wild Bill with his dog pack living there with me. A sense of well being manifest in their presence. He is empowering, liberating.

I met him, through his writing. Just by chance I ran across a chapter of his book *Tales of the Texas Gang*, in a small press effort called *Burning Dogma: The Texas Longhorn Journal of Dada*. His piece was outstanding, marvelous, funny and deep. It was irreverent and earnest.

Then a year later, I encountered this huge bearded guy selling his novel on the Drag in Austin. This is a block of Guadalupe Street across from the campus where they have set up a little market place on a blocked off street that runs into Guadalupe. People set up booths and stalls selling mostly crafts. Wild Bill had a license and a stall and everything.

I stopped, for you never saw anybody selling books there, just memorabilia and psychedelia. I had picked up the book and when I realized what it was I said: "Wow, here's a whole book of his stuff." And bought one immediately. Had him autograph it, which he liked. His book was a big help in my life. I was going through a hard time, working as a tile setter's apprentice, where you spend years doing nothing but mixing mud and grouting down on your knees. It was good to have this irreverent book in my life to make me laugh and feel my own wildness.

We started hanging out: a literary connection. We liked Kerouac. I turned him on to Bukowski. Wild Bill turned heads wherever he went: he was not tall but he was an avid physical culturist and his religious addiction to weight lifting had built a sinewy writhing musculature all around his body. His long beard hung to his chest and made him look like a distinguished man of the old west. Or a refugee from ZZ Top. With long, long sloping shoulders, and defined washboard abs he looked like one of those classical statues of a Greek soldier or a god. In fact he did do some nude modeling in the Art Department at UT for the life drawing classes.

There was a dignified aspect to his personage suggested the idea of a holdover from the Old West, indeed he had that slightly wall-eyed stare those mountain men had in the old daguerreotypes. As though he had just recently come into town out of the wilds and was on his way back. Though he wasn't at all scary, he had grey / yellow eyes like those of the wolf. He would be the first to tell you that he was brother to the wolf. He was just wild. That's why they called him Wild Bill.

Then when I got this place on a tip from Juliet — she too had been a house mate of mine before in another Austin house — and Wild Bill's situation at home got dicey, it worked out for both of us to have him move in. The rent was only $80 a month which we split; this suited our casual laborer / street vendor lifestyle and relieved us from heavy monthly pop allowing us to take four-hour dog walks along the river in slacker Austin bohemia. The dogs were buff and mellow from all the exercise and healthy from snarfing down a whole pigeon they sometimes caught. And drinking the green river water. They loved us and looked upon us with a kind of fond, protective, natural, unspoken inclusion in their pack.

But I am getting ahead of myself. Lets take a quick zoom around the time flower and look at some of the people of Austin and how I came to be here. Of course, the central focal point of Austin is the University. Though some say it is the capitol. But what, you might wonder, do I mean with this concept of time flower. I see my friends as flowers, — petals surrounding their heads, their visage youthful, enjoying their moment in the sun. And upon closer inspection there is the delicate surface of their world which they have organized around themselves like the way a flower organizes its petals to enfold its center. They have allowed me to penetrate this surface and enter their lives. For now I'll just give you an image of it: the people in your tribe are like blossoms on a great branching family tree with roots that tunnel into time,

that is the forking currents of destiny that brought us here. This tribe of friends loomed large in my young life, like peripheral sub-personalities, whose opinion was integral to my way of being. Yes I tend to have this super-ego of smart people around me who can help you get by. The Friends. They themselves often represent uncles and aunts and sisters and good teachers you had in the past. I haven't worked it out formally, Levi-Strauss probably has. Another simple representation would be: The inner child, and the outer masque. Inner and outer articulating gestural movements among the several dimensions of a memory database would be another instantiation of this time flower model. We'll get more deeply into this model, we'll go all around it, even invoke the tesseract for comparison. Allow me to just say that there are moments of liberation to be had from learning to see the unfolding of your life in the time dimension. Of course many have said this.

So then to get back to zooming over this extended space of minds integral to mine own. There is Lenora up in the Avenues, married young Steve, kept a most interesting house, an oasis, you entered by a tall gate, into a groto, with pond and ferns. She taught piano. Her father used to play the cello with Einstein — the great scientist of relativity and avatar of our time, had given her father one of those pictures of his wild hair self — in a black leather jacket. It was signed and dedicated to her father, who said "he liked playing with amateur musicians."

Or further out by the airport, Kenny D. an engineer and musician playing with various bands. Austin had a great music scene. There was Ben, an artist and his brother Robert, a philosophy major and student of Raja Rao. I met them through their sister Martha at a candle light vigil protest ceremony around the time of the invasion of Cambodia and Kent State massacre. We were on the UT campus with the vast throng, on the South Quad just in front of the Main Library, beneath that infamous tower and Kenny M was there too, cupping the

candle with the palm of his hand all night to keep the flame from getting blown out during the vigil.

Then there was the Church of the Coincidental Metaphor. I was part of this group, PS, Brother Human, Craig, John that did border blasting broadcasting on high powered Mexican radio station. They did a parody, wrote zany blasphemous radio skits with Brother Human trying to be crazier than the "real" radio evangelists the station carried — these were Rev. Ike and Bro. Roosevelt Franklin hawking various "prosperity packages." You could hear The Brother Human Hour all up through Texas into the American bread basket, as they raffled off footballs autographed by Jesus Christ, and much, much worse. I got to be a chorus voice in some of the radio skits. We took up a collection and paid $28 for fourteen and a half minutes; sent off the money order to Dallas. I also helped PS with *The Salamander Weekly* a humor sheet we hawked on campus, for "only a penny." Craig, PS and others, Bruce Sterling, decided to run a "ticket" for the President / Vice-President of student government as a lark. They staged all sorts of events of street theater, writing slogans and campaign promises. One was to change the University slogan on the main library from "Ye Shall Know The Truth and It Shall Make You Free" to a more straight-forward "Money Talks". Another campaign idea from this Anarcho-Absurdist party was to have the police cars driving around with only one person in them, to perform community service as cabs. Bruce Sterling was part of it. Their ticket ended up WINNING with the highest voter turnout in about a decade. Yes, the 70s were an ebullient time. The Viet Nam war was ended by popular outcry; the truth got Nixon impeached. This time in Austin showed us how to take that great Texas braggadocio into Absurdest humor, how to take Existentialist black humor into southern soulful black humor, and how to roll it all into a great therapeutic antidote for the oppression and conformism of the times.

Austin was my base, the place I kept returning to because I was unable to escape it. As I mentioned, I had had a little farm outside in Manor, had to hitchhike out there or take my bike, and I got horribly busted while growing pot. That's another story. The bust occurred during a time when I had a job as a night intake worker and guard at the Travis Country Juvenile Detention Center, a long bus ride way out on South Congress. In those days they still picked up runaways along with other underage criminals. I thought it would be good for these youths to encounter someone who wasn't an oppressive authority figure. But you had to be on your toes for youth can react with extreme stupidity. I gracefully recused myself after the bust. So now living back in Austin again, I was trying to clean up my act, hold onto a job, doing casual labor construction work for Neverland Construction or Tao Ono Inc. when it didn't rain.

The one thing the news media and the historians and the scholars missed about the hippies was their conversations. Some of those freewheeling conversations I had with hippies were the first intimate and metaphysical moments I ever had in my life with other people. In my 30th year I was still trying to come down from all that heavy mental Vector Space Theory of Matter, "woo-woo" physics stuff I had studied in college about quarks and the symmetry of the universe, and the method of probability waves and inference for a fundamentally uncertain world. It had been a kind of western enlightenment and it was such a staggeringly beautiful thing to understand the symmetry at the heart of how the 27 dimensional universe unfolds its laws in the instantiation of reality. But I was trying to integrate my feelings and realize that this abstract art is also an attempt at metaphor or objective correlative for how I was feeling at that moment. The excited high speed physics concepts were about trying to have some understanding deep feelings in the presence of this infinite generosity.

A Shuttlebus Named IF

That Spring a young woman named Laura came into my life. She was just a teenager actually, in high school when I met her on the bus. I was sitting on a bench seat in the middle of the one of those free shuttle buses the University runs all over town. They have routes everywhere, and you just get on. I pass for being a university student. The busses are those big long ugly yellow school buses except these are painted white with orange details — the UT colors. The drivers are cool, and competent. This was the Intramural Field bus.

For some reason — coincidence — a fresh-faced young woman with glorious big red hair got on, and since there weren't that many seats available, and I had scooted over so the one beside me was open, she sat down beside me. She had several textbooks in her hands and in a rucksack over her shoulder. She was wearing a white blouse and a red tartan skirt, which did a lot to enhance her red hair image. It made the apparition of her being lustrous. She was tall and had lovely curves. I thought she might be a freshman at UT and I said, "Wow, looks like they're working you pretty hard."

I have been a college teacher before and have great sympathy for the plight of students. She looked at me, and made a face, as if to say, '*not much fun.*'

When she levels those blue green eyes at you, you were immediately pulled by their tractor beams into her orbit, pulled into the shining planet of her pale freckled face. She was so tall that she had to incline her head a little bit forward. It would be hard for her to appear diminutive. But it was her mouth: her white teeth always protruded from her smiling lips. As though her mouth were not big enough to hide her delicious grin.

I must have looked good or sensitive that day for we started talking about a paper she had due in Mythology. She leaned in and whispered low so the driver couldn't hear: "I go to Austin High." So we were both interlopers on the shuttle bus system. She seemed to be distracted, vulnerable, perhaps a bit frightened, though she was a big girl — tall, voluptuous with the smoothest whitest skin. Wouldn't you know she would sit beside the only other person on the bus who was not a student. We introduced ourselves, her name was Laura.

Laura on a passing train I thought. I might be entering a dream through those eyes. A street car named Desire. A shuttle bus named IF.

I told her I used to go to UT. We rode down Lamar, across the river. She looked out the window past me and pointed out the top of her high school down the river. Wild Bill and I usually passed there on our dog walks, and occasionally saw young people doing PE or heading off somewhere in the woods.

She seemed sad. Upset. Harassed. Much later — after we started dating — she told me that her parents were getting a divorce. She was having a hard time with it. I felt like trying to make her laugh. What do you say to high school girls? I did not tell her that I had just quit my job.

I offered to help her with her homework. I said maybe I could write the paper for her. "Why don't you come over and we'll work on that paper. I love to write."

I told her, "Bring a friend, a classmate, so you feel more cool about visiting me."

I told her where I live. Of course, we didn't have a phone.

"OK, I'll be there around four after school gets out," she said.

I smiled, reassuring: "We will have a *study session*."

When I got back to Baylor Street I had to confront what I had done — telling the current employer to take this job and

shove it. I counted the money on hand.

$1.11 One dollar and eleven cents. That's how much money I had when I quit.

I quickly noted down my Book of Hours:
Thurs: 7:30 -12   4.5 hrs
Fri: 7:30 -12   4.5
Mon: 7:30 -5:00 8 hrs
Thurs: 7:30 -12   4.5

I knew I had at least $50 coming, after tax at the end of next week. Less $13 owed to the dentist and $20 owed to the optician. That left $17. Hmmm. And today was only Tuesday. The prospect of the freedom of unemployment was already beginning to have a slight edge to it. Might be the downward path to oblivion.

To take my mind of that dire prospect I did some exercises — to draw the blood out from the center of the body, to flush the exterior. Wild Bill, a monster physical culturist was having an influence on me here. I pressed weight over my head to give myself more muscle mass in the lift, in the shoulders, arms and flanks. Squats to work up huge chest cavity bellows breathing. This, I thought, is what gets the inner man, the inner beast, the Automorph, respiring and oxidizing. Soon it became he, the inner beast that curled the bar (after removing the big weights), then did some sit-ups to work up the stomach muscles, and finally hit the floor for some push-ups for the arms. He turned over on his back on the floor and from the floor pushed his body up backwards in an arch — the Bridge. I am working on the Wheel Asana, he said to himself, which is where one bends over backwards onto his hands, then walks over, and up around again onto the feet. I am only so far along on the Hand Stand.

I was trying to do a little meditating; did some breathing asanas I had learned in Berkeley.

To breathe and stretch one's core again. To settle, to anchor. To breathe in through the nose on a slow count to breathe out through the mouth on a slow count to let the stress go out on the out breath. To breathe in through the nose on a slow count. To maintain the tip of the tongue on alveolar ridge behind the teeth, to pause for a moment in the gap, the lacunae, the bardo between the in-breath and the out-breath, then to breathe out through the mouth on a slow count to let the stress go out on the out breath. To let the thoughts go — though I get carried off on them; to not chastise myself for letting them take me, but try to pull myself back from them — to my self, whatever that is. May I know it. It helps to sing the mantra, the Gayatri.

To think of it is to sing it. Silently hearing it in the mind.

OM.
Burh Buvah Suvaha.
Tat Sa Vituh Vah Rey Nium.
Bargho Deviasah Dhi Mahe.
Diyo Yo Nat Prachodyat.

It is about going back up through the generosity, through space, time and mind, (Burh, Buvah Suvaha), like these were elements in a quincunx, falling from the OM. The mantra is asking us to think by analogy of the Sun, giver of all, to try to think about the archetype represented by the sun, or a star. To see it symbolically: the sun is an archetype of the One; the sun is just one of many, many representations of the omniscient mind, which through local fields and non-local hyper-dimensional tunneling, is able to pervade the universe with agents of generosity so that there could be intelligent beings to know it.

And to ask protection and blessing on this singer here.

Then I started getting ready for my visit from a young student who might want my help on a writing project.

I started some bread rising.

Then Laura and her girlfriend came over.

There they were, these two young women in my kitchen. Her friend in skirt and starched white blouse. Her girlfriend turned out to be Leslie, the daughter of a friend, old John Manchester. John and I used to teach at a technical trade school, Durhams, together in the early 70s, just after I got out of UT. And now here was his daughter turning up at my door. I knew her! I asked about her father. That made the whole scene get immediately relaxed. In my kitchen two lovely girls. Laura the red haired girl with the green eyes is wearing a peasant blouse over blue jeans. She is resplendent with glorious golden red hair and green-eyes shining; beautiful. I tried not to get caught admiring her. She was still a kid, had those girlish plump cheeks and smooth rounded features of a child's face. Not sharp and angular like a grown woman. At one point she passed her hand over her brow, and lifted up her bangs.

"Well you girls are just in time to watch me knead."

I dumped the risen dough out of the pan and started to knead the dough on the board.

"Maybe you can give me pointers on my technique."

The setting sun coming in through the little window over the kitchen sink reached across the room to shine on the two girls sitting on the only chairs at the red formica table with the chrome legs. The girls were acting casual like they were in their own mama's kitchen.

That's the way we like it. People cool and relaxed. But I was a little nervous, anxious to put the girls at ease. I launch forth into material about what I know of mythology from Jung and Roland Barthes about mythology in everyday life. I tell the story of the myth of Einstein's Brain, how people are trying to understand how so much about the universe has changed. I tell them about the struggle of Good and

Evil in the physical theatre of fake TV wrestling. "Myth is all around you." I mentioned Jung and the Flying Saucers as mandalas of a new religion about saviors coming from another dimension. I might have over done it, but I did want to put these young ladies at ease, after all they were brave enough to visit this old boy's pad. Soon there would be the smell of freshly baked bread, saying lovin' from the oven in my kitchen.

Her classmate, who had seen me before at her house, was relaxed to the point of blasé. She started something going about the creation myth of the Bible. "Like in Sunday School," she was saying, "they teach the standard model of How the World was Made." Leslie went into this little-girl routine, reciting memorized catechism: "The world was made by God," she said it in a sweet child-like voice. She paused. "On the first day He made the sun and the moon." She looked wide-eyed to see if we are following. "On the second day He made light and night and then He made creatures that fly and jump in the water. On the fourth day He made animals that live on the land. On the fifth day He made the trees and the flowers. On the sixth day He made a man and a woman and their names were Adam and Eve." She nodded her head as though she were absolutely sure of her truth. "On the seventh day He rested and He said, 'On the seventh day you rest.'"

We all smiled in recognition.

I said, "You know, it feels like the story of creation we just got at University. The Big Bang. The Standard Model. Now there, is a modern creation myth. It starts out: In the beginning a singularity created the Big Bang. Before that there was nothing, not even Space and Time. And that's the first Mystery, for we can not conceive of a world without space and time. But we need to have it this way because of relativity. Space and time are woven into a fabric by matter.

I mentioned Sasquatch or Big Foot and the American Indians, the Bermuda Triangle.

"These may or may not be true."

Trying to make myth relevant to everyday occurrences and distinguish it from legend, I mentioned the Urban Legend of goings on at Lover's Lane. (It might have been an oblique entry into seduction too.) The girls remembered urban legend about the detached hand they had heard.

"What about "cooties," Leslie said. "In the 5th grade the girls thought the boys all had cooties, Don't touch the boys or you'll get cooties. Yeah, and the boys thought the girls had cooties too."

The girls remembered stories they had heard and movies of being on lover's lane and looking for lights or killers on the loose. "Now those are more like legends, right?" Leslie said.

Laura asked, "What's the difference between myths and legends?"

"Hmmm that's a good question," I said. "I am sure some clever college don has figured that out." The two big names in mythology that I know about are Jung and Levi-Strauss.

"Levi-Strauss" quipped Leslie, "you mean the guy who invented blue jeans?"

"No, I mean Levi-Strauss the old French guy who invented Structuralism. He went down to Brazil and spent a bunch of time with the tribes of the Amazon. And he figured out all this stuff about how myth informs the quote 'Savage Mind.' It is like one of my goals in life to understand his theory. But basically he tries to figure out how the mind ratchets itself up on binary opposites. Like how, Men are supposed to be from Mars and women from Venus. (All that hippie bla bla.) Men are supposed to be into war and women into love. Men are supposed to be barbaric and women civilized. Men are competitive and women cooperative. See, looking at things in black and white.

"Oh, yeah," said Laura, "I think our teacher mentioned him in our class. Our teacher is having us look at T.V. commercials. They use a lot of myths. The Jolly Green Giant, the Ajax White Tornado, the Cookie Elves.

"And the Zip Lock Baggie Alligator."

"The Zip Lock Baggie Alligator!"

"Yep."

"Damn, we sure are missing out on a lot not having a TV."

"Yeah, the media department had a TV in the classroom and he had a tape of some TV commercials and we watched them. He said the function of myth was to show you how to live."

Leslie looked bored. "Is this going to be on the test?"

"Yeah, its everywhere," I continued. "Like it gets into Race. Whites are good and blacks are evil. Like that.

"Anyway what I know about Levi-Strauss is this: he arranges all these motifs in myths, in tables and finds the same motifs all over the world. So he concludes they are basic to all human minds. It was supposed to be like music for him. A major modern poet, Ezra Pound did this too with his great ungodly poem *The Cantos*. And that is what Jung is talking about with the Archetypes too.

I could see they were getting bored. "Too bad I don't have much music, so the musical analogy is lost on me."

Leslie piped in: "Laura plays the guitar."

"Oh, cool. You'll have to bring your guitar by sometime."

Trying to be more entertaining, I played the air-guitar: "Strumm, strumm." I continued: "The great modern epic poem of our generation, Gunslinger, mentions Claude Levi-Strauss. The poem is a story about the travels of a 2,000 year-old Gunslinger and his pot-smoking, talking, horse named Claude Levi-Strauss. They are wandering in the West looking for Howard Hughes. There's a pun: Clawed Levi-Strauss, clawed jeans, you know blue jeans, and clawed genes the blueprint that brought you here?"

To pull cover — the gentle old scholar — and put them at ease, I ran the risk of being overly verbose. I continued. "The horse learns to roll these Tampico bombers. There's a hilarious scene where the too-stoned horse named Claude Levi-Strauss, starts explaining to the other *outside* horses that they are not really tied up. This news is beyond the heads of most of them. But one does get it and runs away. The horse named Levi-Strauss tears the saddle of another with his teeth and throws it through the saloon door — knocking it off a hinge. This causes a great quiet pause among the horse-owners inside the establishment.

"It's insanely funny. The guy who wrote it, a poet named Dorn, must have been really high when he wrote it."

The girls looked a little distracted.

"Anyways," I recovered, "to answer the question about legends and myths, I understand they come from the same place, probably they are stories constructed by people to explain and celebrate connections — especially connections that embody stuff that is sensed by everybody, though they might not be able to put their finger on it, but something deeper, more attuned with the times.

"They are a sign for something else."

They looked puzzled.

"You know a *sign*. Like for example if you looked up in the neighborhood and see smoke rising in the air, it is a sign of fire. There is a direct correlation between the signifier — smoke, and the signified — fire. It is immediate. Legends and myths occur in a population and over a longer time. There is some basis —the fire — for the current sign, the myth.

"In some cases the myths are so old, that the referent is lost in time."

Luckily at that moment Wild Bill came in with his dogs. His powerful male presence immediately made the girls get

a little nervous, for he can be a scary looking hombre. They got quiet at the size of the big galoot. Leslie especially. But then they become caught up in the chaos swirling around brought in by the three dogs.

I joked with my friend: "Speaking of Big Foot, we have our own Minotaur right here. Let me introduce you to my friend Wild Bill."

He was like the goat-footed balloon man of Spring, the Pan figure, the natural spirit, resplendent in all that fur.

The girls were charmed by the dogs, for Wild Bill keeps these dogs clean, always giving them baths. And the dogs are like wild natural children in their fuzzy pjs to the girls.

We had a fun time talking about Dracula, the Loch Ness monster and Santa Claus.

Wild Bill was a gentleman easily joining in, telling stories of the little green Reticularians — beings from the star system Zeta -Reticulans. Someone had pointed it out to him in the clear mountain night sky of New Mexico, where he had lived for a time. He mentioned the Roswell cover up.

The school girls in their crisp girl blouses were sharp.

The smell of baking bread wafted through our humble kitchen.

"You girls have to stay around for a slice of bread, it shouldn't be long." I got a mechanical travel clock and wound it up.

Laura mentioned, "Our teacher has us watching commercials on TV. He says commercials are myths."

"Oh, really, I had not thought of that."

"Yeah, like for example, the great struggle of butter over margarine."

"How does that go?"

"Well you know like the insignia on one pack of butter is a crown. So that is supposed to "signify" as you said, that that margarine is the King of butter.

"You see margarine is considered unnatural, or not as natural or wholesome as butter, so they have to get all scientific in their adds, and they have to talk about dieting."

"Wow, who would have though such a deep struggle was going on. Wild Bill and I don't have a TV set, haven't seen a TV in years. Hell, we don't even have a phone. I guess we are out of touch with reality."

Laura made a joke about being a red head. Told about being called Woodpecker. Carrot top. Freckle Face. The light eyelids over her green eyes bat and she flashed a smile, she has a sense of humor!

After a while, after the bread came out of the oven and everyone had engaged in the partaking of big slabs of bread sliced and slathered with butter, the girls gathered up their books and left.

And I, though broke and in need of bread, have been kneading the dough of fresh leavened stuff of life. I feel like Wild Bill and I have created some down home in my kitchen and am filled with the hope of new love.

Even this little bit of entertaining gave me pause, though. All these archetypes hidden behind the electromagnetic emanations. I was driven to be getting back to work on some writing. Existentialism to fill the void. But the void is really a vacuum in detente, in equilibrium with all kinds of potential annihilation and creation going on beneath the emptiness.

I'm working on a theory, but it is hard to think when you are digging ditches or scrambling over house roof frame. I should be "getting about my father's business" as they say. My literary fathers.

It was a big deal to be going 'over 30.' Do you invoke the "Don't trust anyone over 30" rule even on yourself?

I don't have music in my background so the analogy to

music that Levi-Strauss built his thinking on doesn't do a lot for my understanding. Of course I have heard a lot of music and know a lot of tunes, and do appreciate the push and syncopation and fleeting chaotic play in jazz, and know it is like something that takes over and entrains the brain. I do get that music relates to the notes that came before, and flows because of that, a thing pushing itself along by its own gestalt. People recognize that. And the dialectic of theme and variation. I did get the different perspective in Group Theory.

Quincunxial Yantra of the Ancient Sanskrit Gayatri Mantra from the Rig-Veda

Dogmen Down on the River

Whenever we could, Wild Bill and I went on a dog walk down by the river. He was good company, could hold up his end of a conversation, was curious, could be quiet and listen too. As could I. We usually headed out up the hill past the Castle and through Clarksville. On these walks through the narrow unpaved streets of Clarksville, the dog pack stayed close in the narrow parts, though Griz went to the edge of the street. Other times we just went down Baylor into old west Austin beside the river and crossed the railroad tracks and got down on the river that way. Sometimes we'd all pile into Wild Bill's 65 Dodge truck and drive down to the river's edge. He'd park in the Safeway lot down by the married students housing, or at Deep Eddy pool which was right on the river walk. Wild Bill released the dogs from the back of the truck by saying, "OK lets go." Sissy waited for him to let the tail gate down, but Griz and Caief leaped over the sides.

We crossed the river on the footbridge beneath the Mopac freeway. This huge freeway colossus with columns holding up the structure seemed in my imagination to be part of some great temple complex, a portal through which we entered another world. We went down to the edge of the cool green river where it was wide and deep, shimmering green beneath its surface glancing gleams. On the ancient river's edge it was shady underneath great overhanging boughs.

The dogs got busy killing a pigeon that wasn't quick enough to flee. It must have had a broken wing and couldn't fly away. They tore the morsel of wild food to pieces and ate it feathers and all — crunching loudly on the bones.

Typical specific day. Wild Bill talking about Indians,

aboriginals, spirit warriors. Telling his story. Being quiet together walking single file through the dense parts of the forest.

Wild Bill called them his Medicine Dogs. I did not know at the time that the term "medicine" was just vaguely related to the American Indian concept it was supposed to translate. However I did experience the concept through being on these dog walks. When we walked with the wild and joyous dogs in the woods and forests, a bond occurred. It was like they were conducting us in their world. We walked the line and they fanned out to cover area. They were so much faster, to penetrate the space with agile speed, knowing it in a whole adjunct phenomenology, that of scent.

Sometimes it was as if you would look over at the outlier dogs and they would glance back at you and smile. It was a look that said: *Isn't the forest cool? — all this beauty and wildness*. And other times it was kind of teasing look: *Come on! Slow poke*. And other times it was a look of love, as if they were saying: *I'll protect you from anything out here, nothing is going to happen to you on my watch*.

Dogness, we had been taken into the place of Dogness by these emissary beings from the Spirit Realm. Grounded on four legs, fast as all get out, with a proliferation of nerve endings far greater than man's in the long proboscis to plunge them into the phenomenological ocean of smell, which is to say the percolation of matter's vapor trails in space and time.

And so began my induction into Medicine. This term is not a very good one for us blind, materialist westerners — descendents of Cain, expulsed from even the area outside of the garden — to get an insight into the concept. Perhaps "emanations from the spirit realm" would be a better term for it. The dog was our guide. We were like Dante and Virgil entering the landscape of the Real to be received through the only portal available — the mythological; we were like

Cabeza de Vaca or Levi-Strauss entering the world behind language. I was working on a theory or aesthetic of writing that was influenced by Dante and Rablais. I had written a novel titled *Ontological Hysteria and the Austintacious Stomp* about the difficulty of the artist in Austin and about the Church of the Coincidental Metaphor. Its main character was Panurge. I was a Peter Pan kind of guy desirous of following up every urge. As I look back on it now, Peter Pan is like Hermes, or like Michael the Archangel — archetype of the mediator between realms.

When I read Rablais and especially Dante, I got to see through the labyrinthine workings of the medieval mind. It is structured on the three persons in one god analogy of heaven purgatory and the abyss. Rablais also had this structure under his wild improvisatory works. Though that was long ago, I got going on some kind of aesthetic theory bothering those writers then. It was that a work should move on 4 levels: the sensual, the psychological, the symbolic and the mystical. Here is what I took that to mean. The sensual is about creating a good accounting of the physical world for the reader, so that he could be grounded in a present. The psychological is about exploring your own feelings, and the intimacy of sharing yourself with others, letting them know you through your honest confession. The symbolic, I took to be the hypnagogic, that level of direct communication through images when the mind is in the between state, between wake and dream; it is here that one gets a sense of a mind thinking, flashing into spaces near and far, for it is here that the mind leaves the confines of the individual. And the mystic — which I confess I haven't much of a clue, though I imagine it is to get a sense of the Source, to be able to see your pathways over long time as having been guided by some destiny or purpose. (I know that is not a very good take on the mystic.) The mystical life is certainly something I have been interested in. I think of

writing as my spiritual practice to be making myself ready to receive moments of synchronistic illumination. But it is not like I have any kind of control or handle on it.

The term Medicine for the Indians had to do with the idea of dream sharing. Presumably the healer (a shaman) was able to have insight to a problem by dreaming about it, or even seeing into the sufferer's dream. They dreamed big.

This is the Colorado river of Texas that flows through Austin, not the big Colorado river that flows through the Grand Canyon and gives up all its water to the farms and cities of the western lands before trickling into the Baja. This Colorado river that meandered through Austin flowed out of Texas. Wild Bill and I were philosophers, sadhus walking, wandering along its banks in the great adventure of life that is unfolding. With the Medicine Dogs we are drifting into myth.

On the narrow trails, Wild Bill was walking ahead of me, passing through the dapple shadows being projected on our bodies. It was relaxing to be in his presence, there was a respectful distance like how the wild dogs were: they do not do anything back to you so that you can be your true self. He was aware of his heroism. When he walked it was like some great wild animal stalking and opening himself up to the rhythm of the real, what was going on. He was so built-up and muscular that he was busting out of his jeans. And he had a way of relaxing his shoulders — these long, long sloping shoulders, girding like a fortress, the head, which was crowned with wavy hair, his face resplendent with long, dark, full male beard. He was ready and steady, his gait a suppleness that could if need be suddenly explode and twist around and bring a blow up from off the floor that was savagely powerful. It was a relaxed awareness, a sinewy nervousness.

The forest swayed in the gentle warm breeze and shadows danced in the play of sunlight as we walked back into time. Ancient paths — deer trails. Opening out wider flatter. The dogs can fan out, they become like the fingers of a hand, the reach of hunter consciousness into the land, an extended awareness into the immediate vicinity.

Along the river on the dog walks my imagination soared, it becomes a tributary of the Great Kenouchcumandonada River. (The name means long in Cheyenne.) It starts in the mountains of Montana, flows south through Colorado, across New Mexico and Texas and empties in the Gulf, and if flows north through Colorado up into Alberta and Saskatchewan all the way to the Bay of Fundy. Even longer and wilder than her sister the Mississippi, this river cuts the north American continent in half. Along its banks are vast lakes and settlements and great fishing villages.

Drugs take you into myth. Campbell is our guide. And Levi-Strauss. And Barthes. And Castaneda. Wild Bill and I did drugs at every opportunity.

It was great to trip with Wild Bill, he made everyone around him more relaxed. He said, I like acid, peyote/mescaline, psilocybin/mushrooms, natural things. He told me that on peyote he sometimes got impressions of his past lives. He had places in his history, a history much longer than an individual life, that he would return to. I might see him in this: as a sudden body motion of a gunfighter, or in a dodge away from the thrust of a knife. He said, In the world of man there are places in my history where I have done man killing.

Wild Bill often talked of dogs. He'd say things like, Dogs are always in the present moment. They are not troubled with guilt or shame.

Though they do know fear. Like all animals they live with danger. Their world is immediate, they are supremely connected with they senses.

They are in a powerful awareness and are not distracted by image.

I might challenge him with, Well what does the dogs eye tell the dogs brain? Don't they get an image of chow when they hear the dinner bell. What about Pavlov and all that.

He'd retort with something like, It is as though they are extended out of their bodies. For the animals, the world is just as it seems, a continuum that they inhabit. For us it becomes an image that we see. It is hard to come to the world without a whole bunch of thought and machinery.

Then he'd get my enthusiasm up with some of his wild ideas, like this one of Dream Radio: I think the American Indians did have a kind of dream radio with their totems of animals, raven and bear and eagle coming and going in their dreams as in the world outside.

Dream Radio! I liked that one.

For the Indians, their totem animals and the encounters and the objects in their dreams imbued their life with a special connection through the landscape of their dreams. Because these animals were so well connected with the earth by their keen senses of smell and touch and hearing and intuition, it was through the dream that these perceptions were transmitted to the dreamer. It is as though the Indians could extend their own perception beyond the edge of their bodies and out into the world through these animals. The waking world and the dream world were part of a continuum that flowed seamlessly one into the other. They inhabit this continuum, like a hunter knowing his prey, even speaking to it, asking its permission. Thanking it. For us we have lost this real continuity. The world becomes an image that we see.

The forest got thicker as it seemed to pull back from the river and stand around like a proud crowd of beings climbing up a hill from the river.

The river was quiet, just moving along its wide expanse — slow. It made less noise than Waller Creek where it flows through campus rushing down over shoals of rocks on its way into this river.

I imagined we were like Cabeza de Vaca and Levi-Strauss, walking in the new world. Cabez was like a saint, he brought his own time with him. Like Don Quixote, the Knight of the Rueful Countenance, trying to do good in spite of all the setbacks, deprivation, loosing all their clothes and gear, not knowing the language, holding not coin or currency, moving through here, flayed by insects — shedding his skin — and undergoing shamanic transformation, but every day getting back in time for vespers. And Levi-Strauss going into the structure of mind, with his great cross cultural matrix rows and columns transcending time and distance picking up on the zeitgeist of the times from the refugee Jewish mathematicians fleeing Europe, with their matrices — representation of the transcendent symmetry. Cabeza did wander across Texas. He might even have walked through here. Working as a slave for the local Indians, he learned the language and customs.

We saw a wetback encampment. There were 6 young males, primitives among us. We were primitives too. Nodded. Wild Bill must have looked like some 18th century gentleman, or a wild Mountain Man to them, but they accorded him respect under a watchful eye. It was just an encampment of hoboes. (Coming to take our jobs. They could have them as far as I was concerned.)

Ever the mediator, I wax enthusiastic on their fishing equipment. This was fishing line wrapped around a 40 oz. beer can which had been wedged into some rocks by the river. It was a trot line. One took up line fast by hand with this system. Effective for fishing. I used to fish the Guadalupe River over around Gonzales, in a boat, setting out bate on the trot line, then taking lots of perch and trout off the next day.

Cruising along the shore there were the elegant River Nymphs [Dragon Flies]. Red head.

Near the edge. Eyes glancing sideways. I thought about Laura, her comfortable, pretty, rosy-white teenage body of feminine comeliness. Laura. In jeans and boots. She was cool, trying to belong, one of the guys — impossible with that face and those eyes. The wind rifled her hair. She was ambivalent about makeup — didn't need it, on her soft, charming, downy, peach-fizz face, her flushing cheeks. But it was her arresting, vulnerable gaze, penetrating you and yet laughing at you with her eyes, that haunted me now. The wind teased, over her forehead and lifted up her bangs.

I was smitten. I wondered if it was going to go anywhere. I hoped so. She was a sweetie, to bring out the sweetness in me.

Stellae at the Dream Tunnel Portal of the Ancient Future Beneath the Golden Highways of the Sun

Like a Coyote in a Prairie Hail Storm

Wild Bill and I talked about anything and everything. We were surprised to find out that we had both come before the same judge. We discovered this as I was telling him about getting busted on the marijuana farm.

"It was the end of August on the farm in Manor and my partner Gregory and I had already started the harvest and had half a dozen plants hanging upside down, drying in the old tackle shed. We were sitting under the tree on the back patio drinking lemonade when I heard the helicopter hovering wuppa wuppa like some giant mix master —hovering over out back by the pot patch. Suddenly it was just there. I saw it as some giant horrendous mechanical insect looking huge and horrible some 50 feet in the air over the pot patch. The bulging bubble glass of the helicopter was angled down, looking, scrutinizing the pot patch through its huge dark insect eyes.

"Suddenly we knew.

"'Oh no! It's a bust!!'

"I didn't know what to do. We knew we didn't really have a chance to run for it. We'd have to go through thick brush; the laws would have the dogs and helicopters chasing us. There was a good chance we would get shot. All my stuff was here and the place was in my name. The authorities could lock it up and stake it out forever. My heart sunk as the realization set in. The situation was like that old coyote out on the prairie in the middle of a hailstorm, no place to run, no place to hide—this was one of those situations where you just had to lie down, put your paws over your ears and take it. We would just have to get busted.

"The police would be coming up the long driveway soon, so we had only a few minutes. Turned out they cut the lock with bolt cutters. We put Butch and the dobermans inside so they wouldn't get shot. Greg and I took down the half dozen plants we had drying and the little bit that was already manicured, snapped it into shorter lengths and stuffed it into a plastic bag. We knew the police would look all through the house. Gregory took the little stash out to his safe cache—a garbage pail buried under ground, with a board and dirt on top of that, and a stone on top of that.

"Then I started worrying about, 'What should we do to get prepared for jail.'

"Greg said, 'I don't know, I've never been there.'

"'Me, neither.'

"'Well,' I said, 'mom always said to wear clean underwear.' We started putting on clean underwear and socks.

"He said, 'I've heard it's really cold in there with the air conditioning.' So I put on a couple of undershirts too. I stuck my toothbrush in my back pocket.

"Soon came Austin's finest swarming. A DPS deputy, was out of the shoot first with pump shotgun in hand, running alongside their still moving car, yelling, "Hands up!" He leveled the roomsweeper at Greg and I. Two more officers jumped out of their cars. They quickly crouched behind open car doors and pointed enormous .38 service revolvers at us. Out of car three, came two more braves. Very big beefy fellows, wearing stiffly starched uniforms over bullet-proof vests, they were easily closer to 300 pounds than 200.

"What the peace warriors encountered was two nice, long-haired, young, gentleman farmers sitting up straight in old metal lawn chairs on the front porch, their hands in plain view resting on the porch rail. Greg and I tried to look innocent. Our dogs were barking and throwing themselves against the door inside.

"'Turn around and face the wall! Put your hands on the wall! You're under arrest!'

"The dogs went wild. Butch the extremely kind and gentle little border collie had his teeth displayed in a savage snarl. The two dobies, Ada and Hijo were throwing themselves against the walls in the bedroom. Some cops came up on the front porch. From then on, it was all standard procedure. 'Hands up against the wall, feet spread!' The splatter gun moved in closer as they frisked us. A steady stream of peace officers pulled their cars up to the gathering. Greg and I stood with our hands against the wall in the heat of this, the most awful of August dog days, until our feet got tired.

"'IT SURE WOULD BE NICE IF WE COULD SIT DOWN!' bellowed Greg.

"'You better keep your hands up there and your mouth shut,' said the cop.

"Then the main man, who's bust this was, pulled up. As he got out of his car, and walked toward the house, he saw that I had my hat on, and so he went back into the squad car and got his white Stetson. Then, proffering his badge and card toward me he said, "Oh, by the way, I don't know if you've seen one of these." I read the name. Renee Martin, he was part Mexican. Greg and I, in our handcuffs, were seated on the front porch. The cuffs made us lean forward in a very subservient position.

"'Who lives here,' a big mean looking fellow with beard and long hair and scraggly looking van Dyke beard said.

"'I do,' I said.

"They kept coming up the steps. Sheriffs, policemen, constables, deputies and one exceedingly obsequious little chicken shit police follower who said he was a freelance reporter with the Houston Chronicle. I took a particular dislike to him.

"The sun beat down and Greg and I writhed uncomfortably in our seats. It was like an awful assault on my peace. I remem-

bered the advice Malcolm X gave for these situations: You want to try and appear as an equal man. Try to relate to them on a human level and they will try to relate to you. 'There's a big jug of cold water in the fridge,' I said. I was trying to find areas of common experience. Unfortunately the only thing I could think of that we had in common was lawbreaking and marijuana cultivation."

Wild Bill smiled at this

"A cop recited the Miranda from a little laminated card.

"One cop named Mike was really fucked-up. He came up onto the porch with the machete. He sat there with it hacking dully on the wood railing looking menacing, while his coworkers searched the grounds.

"The laws walked us back to the patch. The cops wrestled with the pot plants jerking them back and forth until they wrenched some of them out of the ground. That was too much work so they brought out chain saws and hacked up the beautiful females while Greg and I stood there with our hands handcuffed behind our backs having to watch the massacre. Four big men worked for a couple of hours with machetes hacking. I thought I could hear my unborn babies cry. Greg turned around and looked the other way in disgust.

"The reporter asked. 'Why's he crying?'

"I looked and sure enough Gregory, all 6'2", 175 pounds of him was walking around, shaking his head and crying openly. It was quite a sight. Youth will be crushed. They filled up the whole back of a pick-up truck with pot plants. It was mounded up over the top of the cab and weighing it down.

"They took Greg and me into town and the process began. I remember talking to the cop in the car as we drove in. It felt good to get into the air conditioning. I said, 'Someday they will think back on this as prohibition and see how wrong it was filling our jails with people trying to explore consciousness.'

"He said, 'Yea, you might be right. Could be you're ahead of your time.' He kind of chuckled at the pun.

"We slipped in the air conditioned car through the ranchland past big cylindrical silos of gleaming aluminum full of cattle feed. Passed white rail fences. I put myself on automatic pilot. It was all just too disgusting. At the highway patrol office they moved the handcuffs to the front.

"I remembered a lyric from a Bob Dylan song: 'Walking upside down inside hand-cuffs / What else can you show me.' I needed to make a gesture of defiance, no matter what the consequences. I bent over as if to pick something up, then placed the palms of my hands squarely on the cold marble precinct floor, and even though my hands were held precariously close together by the handcuffs, I managed to do a handstand. I then amazed the assembled police, clerks, secretaries, and anyone else who saw it, to see this tall longhaired man, dangling upside down, hefting his weight up from one hand to the other, moving slowly — in halting, staggering, quivering, balancing moves, across the hallway as I did indeed walk upside down on my hands while in handcuffs."

Wild Bill looked impressed at this.

"We got booked. We got our mug shots taken with a number written on a slate that hung around the neck on a chain. We were put into jail. I was by himself in an ugly cell in the downtown Austin jail.

"The bunk was a solid slab of steel. After a few hours, they put this old drunk cowboy in there with me. The old cowboy went to lie down on the bunk and immediately jumped up and dusted off his duff saying, 'I do believe that is the hardest bunk I have ever sat on.'

"They gave you a blanket. When I looked out through the bars I saw cages going off into the distance as far as the eye could see, tier upon tier of extruded cages, bars aligned in parallel spreading infinitely in all directions, like some great

metal shelving structure molded by society for displaying human specimens. I whiled away a little time scratching mathematical equations in the thick, hideous, peach-colored, multilayered paint covering the metal cell wall. I thought of the mathematician Banach occupying his mind and escaping his pain by inventing discontinuous topological spaces while in a concentration camp during the 2nd World War. It was like being in a submarine or an infernal machine, with all this metal and pipes.

"Gregory's mother, Lenora got us a smart lawyer and we were let out the next day.

"I felt totally devastated. I had no place to stay and only a dime and a penny in my pocket when I got out of jail. I did not want to go back out to the farm ever again but I had to go back several times to settle Butch with the new renters.

"In court for the preliminary hearing, I was surprised to recognize the judge as somebody I had seen around the college campus. I learned his name was Steve Russell."

"Steve Russell!" exclaimed Wild Bill. "He was my judge at a court matter."

"Really."

"It was getting busted for shoplifting food."

"Well, the guy is OK in my book. I'd like to hear about your encounter with him. But first let me finish this story about my bust.

"I *remembered* this judge as a big burly hippie guy at UT, always wearing a raggedy T-shirt and with a big Rasputin beard. As a college student this man had been an organizer for the SDS—Students for a Democratic Society— which back people disparaged as a communist youth organization. And here it turned out that he had gone all through law school and become a judge. Now he looked magnificent. He had the trim beard of an English upper crust lord, and he looked steely-

eyed, in a double-breasted, blue-serge suit under his black robes. And he was presiding over my case.

"The prosecution brought forth the photos—glossy, $8\frac{1}{2}$ x 11 color photos — suitable for framing—taken from the air. I craned my neck to get a look. God, it was so obvious. We had been so stupid! There, in a vast sea of dry, burnt-out weeds, stood a lovely oasis of moist green! Where did I think I was!

"The prosecution said they had taken over 200 plants of good sinsemilla, worth an estimated $400,000.

"The judge looked at the prosecution and shook his head. He picked up one of the pictures and rapped it sharply with the back of his finger nails and said: 'You had to fly beneath 500 feet in order to recognize this is marijuana. Without proper warrant, you violated this man's air space.'

"The judge glanced at the prosecutor with a kind of sympathetic disdain, making his point in a weary tone as if he had tried to teach this lesson many times before.

"'A person ought to be able to expect a reasonable amount of privacy in their own garden. I discharge this case for incorrectly obtained evidence.'

"Gawd! I was so relieved. I could have thrown myself down on his knees and kissed the judge's hand in gratitude!

"I remember thinking as I walked out of court: 'Ah, once again I am free.'"

"Wow," said Wild Bill. "Well, now I'll tell you about my encounter with Judge Steve Russell. It was a little more personal. It happened when I got busted for shoplifting food.

"I hadn't worked in 3 months and was out of food stamps and was living in my truck that winter, trying to sell my book on the drag and I went into the Safeway to get some juice and some pig liver and I stuck a few sticks of that Coon Cheese in the pocket of my big down jacket.

"Coon cheese is a good cheddar without food coloring sold in Safeway, in nice bricks for my down jacket pockets.

"I had about eight months of a two-year probation left on a DWI. It felt nearly over. I've been trying to pay them fifteen dollars a month, which would be easy if I worked, but I have got behind and am supposed to pay twenty or twenty-five a month.

"You know, coon cheese saved me from a felony. Any meat theft in Texas is felony. Alas, the store security at Safeway had been observing me. I had bought a little sack of grapefruit juice and pig liver and was out the door. Two fellows stood in my way. They said they were Safeway Security, one of them reaching in my pocket. I thought they were employees of Safeway. They told me to come on back inside. No. I can't do that, I said to them. They shoved, I thought to brush them off and run around the corner. I would get the truck later, or a friend could. I worried about my probation. I may have put down the two cans of juice, three containers of liver, because the sack held together. My attitude disturbed these men. I could feel they had nowhere near my strength, but my glasses were knocked off, a lens coming out. I stooped to pick it up, to be putting it in my pocket, and deal sonic body punches and run.

"Put the cuffs on 'im, said the one, and the other, a little stocky Chicano, lost temper, none of their arm locks were working and he attempted body punches.

"I was bent over his reach and he hung his hundred and forty pounds on my neck. A third body, fleshy, entered in my opposition, perhaps a Safeway manager, while I slipped about and extended my arm with my glasses piece.

"'I'll break your motherfuckin neck!' grits the highly competitive Chicano cop, locking his forearms in a stranglehold.

"I argued about the glasses.

"He might have hung on my back.

"I went backward into Safeway, starting to be choked. On feeling the arteries being cut off I had to put down the glasses signaling I had given in. I was taken into the back, then handcuffed, arguing, 'you guys are really something, I was only trying to keep my glasses from getting more broken up.

"The Chicano cop filling his report said 'You're an ass hole.' He carried on about my giving them trouble.

"'I wasn't fighting,' I said in disdain. I knew I had to deal with them mentally, sadly. More cops were brought in.

"I was disturbed inwardly, not too rational, worried about my dog, the truck with dope and derringer. At the cop station finally was allowed to call the Hollar Service Center in Clarksville to have my dog and truck picked up.

In my cell I ached for my wife.

"That night everybody was having a drunken time, people were joking like they were in a playground, laughing, bullshitting, mostly blacks and chicanos. I lay down the body, wondered at my coping with the environment year after year. Jail has the stench of prison. I fear prison. Thoughts of breaking bond, getting alias! ID.

"The Judge was Steve Russell, considered the most liberal in Austin. I did not know this, or that he is friends of friends of mine. Early thirties, fat, trimmed beard, wanted to read me the rules and tell me about getting a lawyer. I didn't want a lawyer, wanted to call and borrow and pay fine, bond or whatever, get out of jail.

"Judge Russell explained, 'A lawyer can get me probation instead of a sentence, that a sentence is bad on one's career or job or credit rating.'

"I explained, 'I am a Drag Vender, just selling a novel I've written, hoping to find a publishing contact, so wasn't interested in any of that other stuff.

"'You must be the infamous Wild Bill,' said Judge Russell.

"Judge Russell stated, that a thief is considered an immoral person, that a publishing house wouldn't want to give an immoral person an advance.

"I had to disagree, while wishing to be inoffensive.

" I repeated, 'I just want to get out of there, take care of my dog and things, keep doing what I'm doing, contact a publisher. I am forty years old and not going to change.'

"'Well,' said he. 'I have read your book. You sold me an autographed copy. I also know publishers and have connections. But I'll tell you right now, you might write something else better, but your book is a self-indulgent pile of crap.'

"I smiled.

"There was this other officer, now waiting to take me back downstairs. Consensus was I could get out on the recognizance bond.

"'Look,' I retorted. 'I have hundreds of people in this town who think the book is great.'

"'We have to go,' said the other officer, a nice plainclothesman.

"'You need a hundred thousand to have a best seller,' claimed Russell.

"I claimed 'I could sell a hundred thousand if I had them,' having to go.

"In the hall the plainclothesman, telling me I should get the recognizance bond, be out that day, said it would have been easier if I hadn't given the guys in Safeway so much trouble.

"'Look,' say I, stopping, hang my arms. 'I could've whipped those guys.' I went on further about the glasses.

"He believed me, I believe. Hope he told the fellas.

"Wild Bill can whip 2 or 3 cops. Maybe 3 or 4."

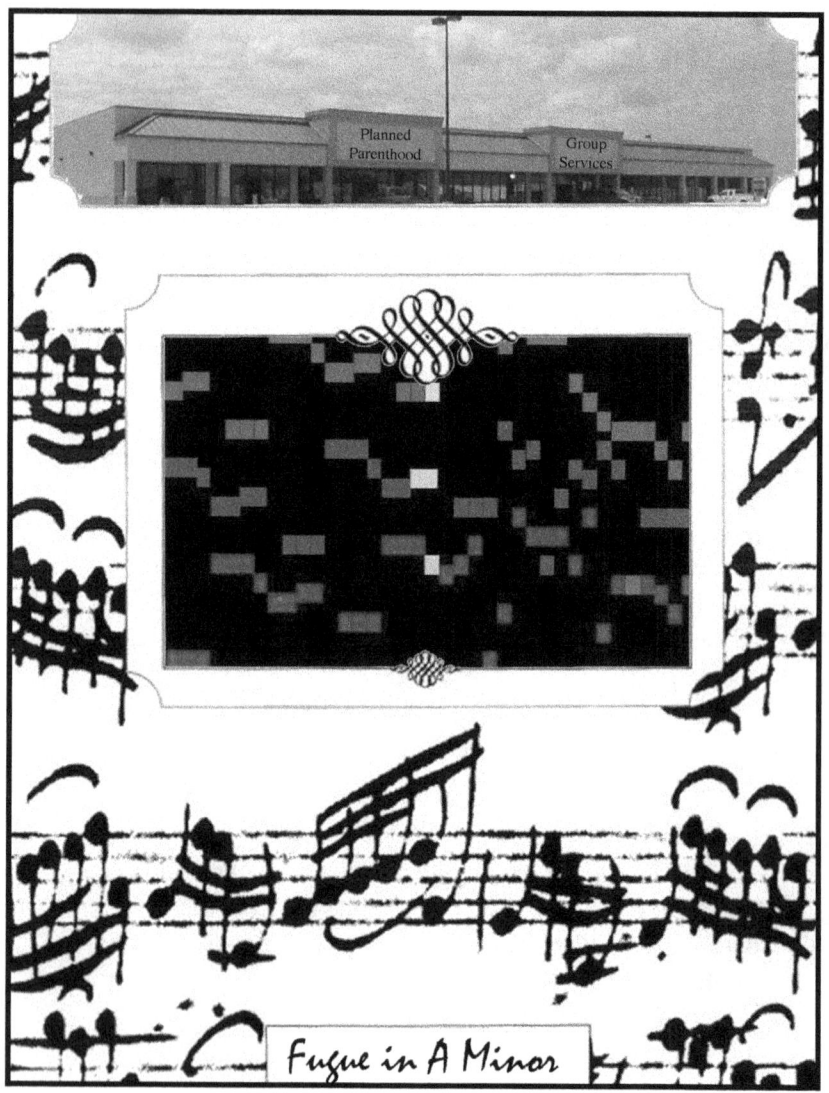
Planned
Parenthood
Group
Services
Fugue in A Minor

Fugue In A Minor

The next time I saw Laura was a few days later. She just showed up at the kitchen door from school.

She was loose, relaxed in loose baggy clothes. Work clothes, blue jeans, work shirt. Austin Slacker type. Tall, fresh faced youth; red hair loose, a teenager.

"Hey, I got an A on my mythology paper!" she said.

She seemed quit overjoyed.

"All right! Congratulations."

I was pleased for her. "Well, we must celebrate!"

There we were, a man and a young woman walking down Baylor St. to the Whole Foods on Lamar, just a couple of blocks away. The big store had just opened. Whole Foods was an Austin hippie favorite, they used to have a store in a Victorian on Rio Grande, when it was called Saferway. It was the place to go for delicious organic goodies. I bought a bottle of Shiner Bock for myself and just an ice cream for Laura.

I teased her a bit: "It wouldn't do to contribute to underage drinking in public by a minor."

She took the teasing in good grace, demurely contorting her face into a mock look of disappointment. The girl began working some of her powerful magic back at me. I could see where it would be hard for one to withstand the full force of her slightly shocked, slightly sad gaze when Laura inveighs upon you with it.

I was hoping she would come back to the flat. I had been thinking about her a lot, wondering, hoping that there was some possibility for us getting together. I had spruced up the pad, wanting to make the environment pleasant, to support seduction.

My humble bedroom has only a mattress on the floor, upon which I had put freshly laundered sheets, and a homey cozy comforter. The room was all wood, unpainted. The walls were big slabs of pine, laid at an angle, very nice. Light hardwood floors of amber maple. The window trim a darker mahogany stain. The room was open with great light: it had a bank of three windows facing south, three more to the west —there were big oak trees with their dark bark outside those windows. And to the east was the one big window, 6 foot tall, facing down town. It was through this window one exited outside onto the deck. I had built a set of stairs up to that window. I had built the deck just outside, resting on the slanted roof. From the top of the stairs, you had to do a low crouch, stepping with one foot over the window sill, crouching and gliding low through the open window, then stand up again on the deck outside. Out there on the deck I had my pot plants growing in green 5-gallon pickle buckets from a burger stand. The deck was hidden behind some wisteria vines that climbed up the side of the house, and up the roof and up over my deck railing. It was a sweet little patch. I was a husbandman of Mary Jane. Yes, I had not learned my lesson from the bust, less than 6 month ago, terrible as it was. I was a recidivist, OK?

In the bedroom my only chair was positioned in front of a homemade desk. I had built the desk too, out of some big 2x12s I cut on the table saw at the construction job only a few blocks away at some condos on 6th and Wood Street, right next to Shoal Creek. I carried home the lumber for this built-in desk on my shoulders through the neighborhood piece by piece. The desk was yet unfinished; I'm not a sand man. My trusty mechanical Underwood typewriter crouches on the desk next to piles of paper. My mind gets projected through fingers in touch with its keys as they click quiet and it pulls words in a neat row till the end of its reach. Then I return. It made me feel fulfilled to bang away on that humble old artificer, under the

influence of the Real. With just one chair it was home. That was about all.

When Laura and I got back I invited her up to check out my deck. You had to go through the bedroom window to get out to the secret deck. I showed her how to stand with one foot on the landing of the stairs that goes up to the window, and plant the other foot out the window on the deck, then do a low squatting slide out the window, and onto the deck.

I was nervous. It was a bit uncool of me, to show my pot patch to a stranger. I reasoned that exposing my secret to her like that, putting my safety in her hands, being vulnerable, would somehow be empowering for her. Or at least make things a little more even — after all here was this teenage girl visiting a bachelor's apartment.

Laura and I sat in the sunshine amid the glorious marijuana plants. We shared a joint. She was totally cool with that. We talked. Being stoned makes me a little paranoid, imagining what was going through her mind. Was she taken aback at the meagerness of my situation?. . . As a hopeful suitor of this young woman I was trying to make myself presentable, desirable, non-threatening, as well as cheerful, brave, clean and reverent. I tried to look at my humble abode from her point of view, as to its suitability as a love nest — her thoughts checking out the room: I was slightly taken aback when I first saw his room. Walker only had one chair, it was at the homemade desk by the bedroom window. The desk was just a big plank. He told me he had brought it from the construction site where he worked. The only other piece of furniture in the room was the bed. He sat on the bed. I sat down on the man's chair. We were stoned and swooning.

I am going to shift into the 3rd person omniscient point of view here, to be that all seeing eye because the story — parts of which she told me, and her friend Wendy told

me are fitted together in a memory gestalt in my mind and also because it is all so sweet and cute, and it is important to put yourself in someone else's shoes to be better able to empathize with what they are undergoing.

He came over and knelt in front of her, in front of her knees. He looked up at her. And she opened her legs for him to get closer in.

And somehow, with him on his knees before her looking at her with those sky blue eyes, it just happened.

Their first kiss. He turned his lips toward hers. She met him half way. His lips were soft and gentle, and wet. She was surprised.

It made her feel so grown up. She pulled him to her, he pulled her to him.

As she explained it to herself: Our first kiss just happened. Not in a rush; nor was it altogether slow, but watching him be so gentle with me, so loving, it just felt nice.

They were both smiling.

"Wow," he said. Like some effervescent bubble had just popped for them.

She was excited, congratulating herself: We didn't go long enough for anything incredible to happen. It was beautiful. Just simple. And I love him. Forever; eternal.

I told him I had to get back home to take care of my sister. He was a real gentleman. He walked me home. Once he took my hand on the streets of Austin. My feet barely touched the ground. After that first kiss. I knew. Sometimes you just know. He might be the one. Even at such a young age. Most people just can't fathom the fact that teenagers even know what love is, but believe me, they know.

"My Love." She starts thinking of him as "My Love." He didn't walk her all the way home, stood at the end of her block and watched her go in. He didn't want to encounter her father who was just a little older than he was.

The next day at school, she told her girlfriend Wendy about it. "God we almost did it last night!"

"Yeah?!"

"Luckily I had to go baby-sit my sister or I would have jumped on that man, that sweet man, right then and there."

"How old is he?"

"He's about 30."

"Well good for you."

"I like older men."

Wendy said, "You KNOW I do! As far back as probably age twelve, I've been attracted to older men."

"Me too."

Wendy was a very mature 17 years old. She is a nordic blond, tanned, warrior-woman type. She is a lot of fun, sly smiles, verbal, jokes — though there was a sadness in her too. She did carry a lot of responsibility for one so young. Wendy had her own car, she drives; Laura not yet. She lived with her father who was a doctor, a dentist, and she even had a part time job in his office, as a hygienist, or assistant nurse. She was always around adults. Her father was gay, — in Texas — still quite a worrisome proposition. Laura was delighted that her friend was pleased. She knew her friend was glad to see her coming around to this way of thinking too.

"High school boys are so immature!" Laura said.

"I know! Wendy agreed. "If I have to spend one more minute around them hemming and hawing and being all incommunicado and mumbling and being all sarcastic and shit, I'll scream."

Laura said, "I promised myself not to date anyone in high school, having just gotten out of a bad relationship with a certain high school boy — named Justin! I am now doubly resolved about that."

Wendy took Laura to a planned parenthood clinic that very same day. It was a free clinic and they didn't have to give their names. It was a glass storefront in a strip mall with a big sign on it. Planned Parenthood. And in smaller text beneath: Approved Drug and Alcohol Testing Programs.

Inside the blinds are drawn to look out but not in. It is mostly Mexican women with kids sitting around in chairs. They were there to get birth control. The people who ran the place were actually very nice. The AA programs were at night. Laura picked up a bunch of brochures on sexually transmitted diseases. She picked up several color-coded brochures about birth control from a rack with titles like:

Your Guide to Getting Started on Birth Control

The Pill brochure — When you choose the pill

The Condom brochure — When You Choose the Condom —Learn how to use a condom, the Condom Do's and Don'ts, side effects, benefits for using one, what to do if it breaks and more.

The Ring brochure — When you choose the ring

The IUD brochure — When you choose the IUD Learn how to get started with the IUD, benefits and disadvantages, and common questions.

The two girls were sitting in chairs in the waiting room of the Planned Parenthood clinic talking and making fun of these brochures.

"Learn about the contraceptive ring, how to insert and remove the ring, the benefits and disadvantages of using the ring and more."

Laura said: "Walker is kind, caring, loving, and very respectful. He's from Canada."

Wendy said, "The boys our age are too immature so I think that's why us women have this attraction towards older men."

Wendy is smiling and agreeing with her friend who

is becoming a woman. "I don't find a single guy my age attractive at all and I haven't in forever."

Laura: He's also incredibly good-looking as well!

Wendy: I just can't date guys my own age or younger, I'm ALWAYS attracted to older men, I think its because I'm rather inexperienced so therefore love to be in the hands of someone much more experienced than me because I feel protected.

Laura: I think its partly because girls seem to be a more mature than males of the same age.

Wendy: Older men are cool cause they have their own place. Seriously, I'm 17 and since I was 14, I have started developing crushes on guys in their 30s and 40s and I thought something was wrong with me! I felt like certain older men I've encountered in my life have truly connected with me and who I am — rather than what I can physically offer. It shows not only genuine interest but respect. I've had a huge crush on a co-worker who's in their 30s and he is attractive, interesting.

Laura: Well if I plan to seduce Walker, then I will probably have to make the first move.

Wendy: I know I shouldn't be considering a crush on this man but I cant help it. He's just so damn sweet to me and respectful. He's healthy and way over 6 feet of pure masculine unadulterated lust and there's no way I am not going to get into some kind of relationship with him, whatever it takes. I haven't a clue how I got into this mess but I really think I love the guy."

Walker and Laura had planned to get together but she got stuck with baby-sitting her little sister. Laura yelled at her mom and got grounded. So it was a while till they saw each other again.

At school Laura tells Wendy, "I've had dreams where I've had sex with him. And then when I wake up the next

morning I'm really horny. Perfect for going to school. And I dunno what to do about it."

That weekend Laura showed up at Walker's flat. She has been coached by the more experienced Wendy, her confidant in these matters of the heart.

Laura was wearing her hair up, to look older more sophisticated than her age. She wore a short skirt, and a lovely colorful red and brown blouse with puffy shoulders and a bare midriff showing tawny tony tummy above good wide hips. Her lovely lips were outlined in red. She looked hot. She smiled a lot, her shy sympathetic smile, as though she understood her effect on men. She had little earrings and come hither eyes. She had a look in her eyes that seemed to say, I know I'm gorgeous and I am enjoying you looking at me. She did not let her hair down.

Walker and the young sophisticate Laura, went out on the deck to smoke some reefer. He offered his hand to help her out the window and she took it. He enjoyed looking at her strong shapely leg — stretching out through the window. They seemed to be touching each other often. Walker was trying not to be caught looking at her fine curvaceous girl gams. When they came back into his room, he sat on the bed and she sat on the chair.

The skirt was hiked up to past her knees.

She reached into her back pack and pulled a whole bunch of those colored brochures she got from the Planned Parenthood clinic about contraception.

"Wendy took me to a planned parenthood clinic," she said. "It was a free clinic and we didn't have to give our names. She's cool. Her father is a dentist, and she works as a hygienist in his office."

She fanned out the colored brochures like they were a deck of cards. She proffers them to Walker. He looks at them seriously, though he is about to burst out laughing. Bless her

heart. He leafs through the sheaths of multicolored fold-over brochures. He had to maintain respect for the forthrightness and the sweetness and greenness of the moment. He read the titles of each colored brochure with concern for the seriousness with which she was responsibly furthering this next step in their romance. He thought, My, she is being very mature about this. With a face as serious as he could muster, to cover up the joy in his heart at the night's prospects, he said: "Well, I am totally OK with the condom. It is, a man's responsibility."

And then he laid back across the bed. He was smiling his little catfish smile at her.

Laura knew she would have to make the first move because he would hesitate considering her age. And then she walked over to him and she straddled him so that he could look up her little skirt. Then she lowered herself down and sat on top of him. Nothing like the direct approach. She leaned in and kissed him on the mouth and they started making out. And since she was only wearing thin little panties under the dress, she began to feel his cock growing hard under her. She started to get nervous and she could feel her pussy starting to tingle below.

She thought: I knew he could sense my excitement, and I could feel his. I laid on top of him and he held me to him and we kissed and I almost fainted — he put his whole being into a kiss. Her lips were soft and pliant and responsive, smooth and lipid.

Later she would tell Wendy, "For me it was being with an older man that turned me on. And for him it must have been because I was so young. Walker wasn't wearing any underpants under his jeans, and his cock was defined against the faded blue denim. I could tell it was a big man's cock not a boy's.

Laura felt butterflies in her stomach. Then she pushed

herself up off him and pulled her shirt off, she pulled the sleeves down off her shoulders, and looked at him with those killer eyes, enhanced with a bit of shadow for the evening's vamp. She lifted the shirt over her head. She reached behind and took off her bra, remembering to straightening her posture. She told herself: *Don't hunch your shoulders. And just be proud of your boobs.*

She got off of him and stood up over him. He sat up and he reached his hand up under her skirt and touched her panties. With his other hand he started unbuttoning his shirt. She helped him pull his shirt off. Then he lay back down and ootched out of his jeans. She gracefully stepped out of her own panties. And stood there looking down at him. He just laid back on the bed there waiting for her, with his cock sticking up. She was so horny now. She wasn't scared.

He said, "Here, let me put the condom on."

"Here use one of mine," she said, and got one out of a pocket in her back pack.

She got down beside him and they started making out hot and heavy. He got on top of her and put his knee up into her crotch. Then he slowly started rubbing his cock up and down her slit. He could feel her whole body was getting so hot feeling it touch her. She could feel the cock's heat and how hard it was. She was getting wet. He wanted to make her dissolve in his hands.

After the head was wet with her juices he slowly pushed it in, which made her gasp. Slowly he began fucking her, pricking her with just the head of his cock.

Laura could feel her pussy being forced to open for his cock and she could feel how tight her pussy was around him. Then with one motion he slowly pushed his cock into her. She could feel her pussy stretching around his cock as it entered her. It kept filling her until she couldn't take any more. He stopped when his cock was in all the way.

She could feel his balls resting against her. Then he started thrusting into her taking his cock all the way out and then letting it sink back into her.

He started moaning softly at first and then he began to get so loud Laura thought the neighbors would hear. The moaning turned her on knowing how good it must feel for him.

Then he slowly pulled his cock out and they both sat for a while breathing hard.

"Why don't you get on top," he said. "That way you could control it. I don't want to be hurting you by going too deep."

Later she would tell Wendy, "Well, I rode him. Which I thought I'd never have the guts to do the first time and I loved looking at him when he came. I didn't cum but that is perfectly fine because you don't usually your first time, and I still got a lot of pleasure. We then cuddled for ages. I even cried a little."

Then they put their clothes back on and he walked her back to her house.

On the way back I was wondering about what I had gotten myself into. It's a big responsibility with a young inexperienced girl. Where is this gonna go, how is this gonna play out. Maybe she has psychological issues with her father.

What girl doesn't?

I was delighted that this 17 year old girl was attracted to me. I will treat her like a queen and be very affectionate, loving and supportive of her.

How did I get to be lucky enough to have a relationship with this teenage girl.

Don't look a gift horse in the mouth! man. If this girl wants to get into your bed, who am I not to give them what they want! What is the age of consent here in Texas anyway? Jesus's mother was like 14 when she got married. They all were, back then.

Squirrel Cat Lookouts on the Great Arboreal Highways of the Emerald Forest

A Good Cry for Green Spring

Vera, the perfumed lady, came to visit Talbot. She was his girlfriend. I had seen her up there on his deck above my place, trimming his hair and laughing. Talbot, the owner of Neverland Construction Company, had hired me to help him on a big children's playground project at the Battered Women's Shelter. He was our resident wizard and general inspiration around the complex at Baylor St. He had engineered the amazing three-tiered tree house connected to the second deck he had built jutting out of his little place immediately above mine. To work with him was like being an apprentice stone mason and tile setter with Gaudi. We did sand boxes and tree forts and rope climbing structures and coy ponds with big slabs of stone for a bridge. He was the one who lead the way at modifying our housing with skill saw, hammer and trowel to shape it into anything we fancied, (forgoing the low $50 property deposit.)

I was on my deck on the other side of the big house from his. I built the deck on the roof, coming out of the east window of my room. I heard his dog Buppie bark, and was sensitive to all that was around me. The dogs do not let any strangers pass unchallenged. I skittered up the slant of my roof, and got onto the ridge peak at the top of the house over my room, and saw Vera come up to Talbot's back porch. It was a warm, misting spring night. The air was so electric with possibility it tingled. I had just managed to mush and gag down, through nibbling and quaffing, some slices of peyote. I was just starting to feel its satyral, sartorial, satirical, intimations circulating like a brewing mystical storm on the periphery of my extended field. It must have been amplifying the scent sensitivities, for like a wolf I was picking up perfume: Vera leaves a scent of perfume, a swath of scent, wide and obvious to follow.

Peering through the branches of a great hackberry tree whose boughs spanned across the distance separating our houses, and feeling like some kind of sensitive animal looking through the bushes at people, I hailed her; we hailed each other.

She talked hurriedly of all the perfumes she had bought in Houston. She is a speed rapper, 26 years old, a style left over from the 60s. She looks like a 50s sex goddess, nice generous curves, wearing tight pants, that she had tied around her ankles, in the style of what the women in France call '*le cigarette*'. She tried to waft some of the perfume smells over to me, by pursing her lips and blowing across the bottle. Then she invited me over. I came across the roof, stepping onto a convenient bough of the hackberry tree, moved hand over hand stepping deftly through the limbs, moving across the space over the ground far below and climbed across onto the back deck of Talbot's place. It comes into the east side of his house, off the bathroom. There are further treadways and rope bridges leading to the other deck and into the treehouse.

There we were in the bathroom together and she drops her trousers and sits down to pee. I got kind of flustered and walked out into the bedroom. She started talking of her desire to explore the psychology of scent. We sat on the floor in the living room of Talbot's house and smelled all kinds of heavy perfumes, in the misty, night air that had gone positively effervescent with humidity. I got those scents all over me, in my wrists, and mustache hair. She even put some stuff on my lips. The two of us must have smelled a fright, like we rolled in it, like musk monsters escaped from an oleaginous explosion in an olfactory factory.

The phone rang, it was Talbot, saying that they were videotaping him at the studio above Folk Toy. We decided to walk over across Shoal Creek and take Buppie. And to pick flowers along the way, she brought a little pairing knife. I

went along because she invited me and I felt protective — for the way she was engaging in such high speed rapping, she might be high. She was good company. I wanted to be with her.

Down on the creek, away from the cars of Lamar, the breeze was cooler and carried the delicate aromas of the sweet-smelling flowers drifting through the air. And Vera could recognize the scent and name the flowers and sniff out the perfume trail back to its source. Indicating a Chinaberry tree, she said "Do you smell a fragrance like honeysuckle coming from that." Or, "Smell that lavender coming from over there by the rock on the edge of the creek." We clipped a whole bouquet of flowers and garlands, purple oleanders, and some white vines, some jasmine, little helpless jazzmen playing in the breeze down on Shoal Creek. I was like some kind of factotem of this nature girl here. I didn't realize until she showed me where, how many flowering trees and cat tail reeds and star-shaped flowers there were blooming underneath the high tree canopy of native cedar elms and live oaks.

I had got the eucharistic spirit, had partaken in the flesh of the gods and the effect was powerful. Patterns and shadows were moving, color escaping their objects and flitting freely through the firmament. And I was walking with Vera, and the moon was full and the wind was riffling through the reeds down by the creek and bits of seeds and flotsam were lofting and I got this image of some Being passing through —maybe it was a female, maybe male — with long flowing clothes, loose clothes like delicate leaves, trailing flower petals in its wake. Yes! The Being wore a cloak of leaves and flower petals. I stepped back and tried to notice the spirit of place here in the draw of the creek bed — so different from the cyborg world of the seething city beyond the green field of the park. I seemed to recognize something — it was Spring! Coming around again. Yes! It was Spring, moving, through the air. And it moved on sweet scent.

I wanted to engage in communication with her: Spring! You are coming back! After such a long hard winter. How I've missed you.

It was like that picture, the Birth of Venus, Venus on the half-shell, with that eternally beautiful slightly sad, definitely stoned-out, long-hair hippie-chick with the braids twisting endearingly down to her shoulders and she's naked and so hot and so beautiful, and the sea waves are circulating behind across the horizon behind her. And the other angel creatures are flying in and the man is blowing flower petals through the air and this absolute babe is hanging on him and over on the other side of Venus is another beauty, running up to Venus going to throw a veil around her and the flowing of the clothes and the dipping of the branches and the flying flowers are all just such a gorgeous, dreamlike wavering in the surface of the veil of the illusion of "reality" before our eyes. This symbol of spring and her associated deities — one trying to get her clothes off and the other trying to modestly cover her, a tug of desire and rejection back and forth like the resilient trees swaying back and forth in the buoyant breeze.

You know that one. That girl with eyes as blue as the sky, looking out at you kind of glassy-eyed, cause she's preoccupied or maybe kind of sad, or overworked, or just stoned — cold from standing there nude like that. You'll start to notice flying cunt-flowers blowing in the breeze and cat-tails standing up, up for her. And as the masculine tries to access her creativity, the feminine rallies around her, to protect her. See the world in a flow: little waves and rivulets and torrents and deluges that come and go. You have to keep an eye out for her. She's easier to see in the movement like how the invisible wind moves against something: thus you will feel the impression of her flow.

Yes the being wore a vestment of flower petals, and it moved on sweet scent. It was the opposite of the cyborg

which was the hard society. And it shimmered through and up against living things, and you could see it sometimes if you trained yourself. You had to get yourself into that state between waking and sleeping. That's where it was. I don't know the brain chemistry or the alpha rhythms — where the thing from the ordinary world of consciousness let go, and consciousness was no longer focused just in the brain, but became focused in the body. The seat of consciousness dropped down into the body, that's what happens in sleep, in REM sleep: the thing, that kept consciousness confined and localized to the brain, opened up, and let consciousness flow back and forth into and out of the body. It was a two way street. And you could go out of the body and float down that street.

You can't see her with your regular everyday eyes, but she's right there in front of you. Always is. Every moment waking and sleeping. She will appear to you on her own time.

She's gentle — now, but always being born and always things dying. It was coming on strong, staggering me in the breeze, a breeze that could have knocked me over like a feather. I wanted to grab Vera and say something to make her understand. Like this: *You've known her since you were born. She's your real mother, you can follow her, her perfume, wafting on the meandering zephyr through the weeds. Feel her, care for her, come to know the generosity, the infinitely deep love and attraction in an infinitely deep universe. All moving — flowing along — toward some other state of purpose. And the attraction which feels out into everything, with field fibers, to draw out from the darkness, from the ground and take into the light — life to be born, loved, each other to be a flaring into light. We were born by this force propelling us into the day of this very blue sky and the night of these infinite stars, that the sun occludes with light in the day, and into which it dissolves in the night.*

And more, in our attraction for the other, each pulling the

other closer to each other, we come to know the pull of love, the lift, the pressure, the push of love moving through. The world, in the intimate kiss from the other, reminds us of the immense love, that the universe has for us.

At the Folk Toy studio space, there were many beautiful people smiling, milling around. The hippie women there compared well with Botticelli's Venus. Even though her face is so beautiful, the faces of these people here now are even more beautiful for they are real and touched with passing Springs.

A video tape was being shot, in color, of a shawl that Talbot had made. That night we saw the architect as tailor. The assiduous perspicacity of his bead work is amazing. This shawl weighed over 17 pounds. It was entirely constructed of beads strung on strong monofilament dacron thread. It took him over 1700 hours to bead up. He had sold it to an American Indian woman in Houston for $2,500. It had a red border around it with Navaho cuneiforms, flat gestures of angular movement that reflected a yantra, in a radiant field.

I tried it on, and looked at myself in the mirror, whirling so that the cosmic garment flowed out. My mind was working, seeing. The shawl of lights, was influenced by magical wampum weaves and integrated circuit design. The strange yantra symmetry of the inter-penetrating triangles of the cosmic eye, that pulled you in, was done in those bright Huichol colors. It seemed to me that in wampum circuits was a codification of the records of the migration of our tribesmen, the indigenous tribesmen of neverland.

Talbot in faded, paint-spattered rag-fragged blue jeans, wrapped the shawl on his bare torso. He has long strawberry blonde hair and his long ginger beard has colored beads braided in it. He is video taped swirling around in the shawl of lights, in front of a shiny deep-blue wall. Blue wall, swirl of shawl of lights. Talbot with his dog, Buppie, were like the boy Buster Brown and his little dog. Buppie is actually a Boston

Terrier. Buppie grabs onto Talbot's pant leg and they go into their dance: he twirls around and around on one leg with the other leg extended, the dog hanging onto it with his teeth. It was quite mad. He loves that dog. He is the architect as tailor, has tremendous energy to get things done. Sees whole pattern language in which he makes buildings to suit people like a tailor might make clothes for people. Take a collar from here, a yoke from there, and fit these together to make a tremendous *hombre mariposa* jacket. Working for him I had come to appreciate how his architecture work is his jewelry work writ large. Like how Gaudi puts smashed plates and tiles into his free flowing concrete forms, or the way the Germans or the Scots do stone work, or the way Mexican tile has flowered, or how to weave in the bamboo and teak structures of small pagodas and spirit houses. Or in the way Texas rural architecture takes advantage of orientation in how to construct a breezeway to use convection cooling, or use thermal mass from underground for heating and cooling. And especially in the intricate tree houses he built in the trees above our enclave.

Though everybody at the party acted all cool and indifferent about it, we all wanted to look at ourselves on the TV. There was some tres chic and stylish ladies. Vera rapped to them about the Alexander technique — where you get your spine erect to cradle the head up and the rest of the body follows. At one point the matron who ran the place noticed that a whole bunch of guys were standing around Vera holding forth, and the lady said we were like bees buzzing around a honey pot. The things these old wives say, made me crazy. After a while they all left the studio to go to some other thing. But Vera didn't want to stay, and I was getting all flushed and rushy so was glad to go home with the one who brung me. It was a hot humid Texas night and we stopped at the Texaco station to get a coke, cause I was dying of thirst. The fluorescent lights were bright and frightening.

Then we were safely up into the center of our place at Baylor Street. And something remarkable happened. She pulled up some long slender shoots of tender young grass. She broke them in half by bending them over and pinching them, then she held them up to her nose in her cupped hands and inhaled them. She held her hands up to my nose, said, "Here, smell. Inhale the perfume of photosynthesis." And I did. The perfuminos, particles of photosynthesis, foam emanating through time fields in Spring's garden. That sets me off a train of images of green dreams.

She said, "It smells so wonderful, it makes me want to cry. I haven't had a good Spring cry in a long time." She looked at me, felt the need to explain: ". . . a cry for all the small animals and the seeds and reproduction going on, all the things coming up, coming alive."

Seeing her do that I was really touched. Holding up her cupped hands, the broken-open blade of tall grass, scent like a whistle. How do these women get to be so in touch; so in the picture. They can have such a casual grace, carry magic and beauty. She lets the shoots fall out of her grasp. I see them drift — flotsam, seeds, food for the next level, recycled by bird in nest or worm or microbe finer and finer, like the pores of a skin, the world is a sponge soaking everything within.

The field that I am extended out into, with my mind and with my eyes and with my ears was coming closer to me to be reached by touch. I saw the hands as finer extensions of the arms, the arms as finer extension of the shoulder, the fingers on the hands as little hands, getting closer and closer, the skin on the fingers becoming the eyes of touch. There were two bubbles touching — me inside one, the world in the other. Where they touched it was undulating and rippling: these water bubbles at the interface of two media, or different dimensions. It was like when the sea foams as it washes up on the land, or where the geosphere reaches through the ecosphere and

touches up against atmosphere, (different world) (dispersion, diffusion, diffraction).

I should have held her then. But didn't.

For this was Talbot's girl. But I'm sure he would have understood. It was my being so heavily imbued in the economic space (he was my employer.) Rummy-eyed, crying on a misty peyote night — green dreams filtered through green light. I started talking to her instead. That for me in these situations is a way of touching too, but also perhaps a way of putting off, a wave of hands in space. A little screech owl who hunts by night flies by screeeeing up the hill toward the castle where squirrels scamper and cats slink and insatiable weasels are out there in the night; but in the country of the night, it is the cat with the infrared eyes who is king. I am.

Who am I in all this? Since I have access to the symbolic I might try to think about being. I'm the angel in the painting, carrying beautiful Laura on the wing.

Zephyr and Aura are the graces covering her nudity.

I am The Spring, the zephyr, who is pursuing Venus.

I am you, the viewer, the man trying to see behind the arras, the veil.

The feeling of Spring and the spirit Mescalito was enough to make me feel these rushy sensations, intimations, condensations from many worlds, all the while having to maintain.

I'm the zephyr, a satyr, an aleatory lyre, the breath of the shokatochi. . .

Vera and I ended up in Talbot's swing, way up in the tree house, sitting side by side on the bench swing, swigging the huge bottle of coke. It was such a relief to be up in the tree house gliding with the breeze up there tickling the leaves. It is so nice to get away from these cars, these behemoth, hunchbacked metal beetles roaring and charging around carrying death on their grill, and loathing in their bonnets.

Zephyr is chaos, and Venus is biological order; zephyr

blows the seeds around on currents, but when they find the right sight, Venus applies her attraction and her creativity and life is born again. She is borne by the sea on a half shell for she is born of the union of a god and the sea.

And I could feel this dance has been going on across the universe forever and you are privileged to be a small part of it for a time. You are a seed, a spark wafted on the wind. You may ignite, and get something started and grow, and you may not.

Vera talked of her desire to explore the psychology of scent. She wanted to study orthomolecular therapy for schizophrenics.

She said, "The person who used to have the flat you have now is one. They recall him to the hospital for bad head regularly."

I recalled his incredible verbal performances in my kitchen. The flat I have, used to be his for many many years during the entire 70s; I occasionally get his mail.

One of his favorites riffs is controlling his movement like a dancer by his firing the retro-rockets. Coming out of his elbows and knees, making him correct his movement like a space ship landing, where he shouts CRAB WAFFLE HEDGE.

She recalled how one time he came into the drug store where she works as a pharmacist, and with these wild eyes shouted out, "When you die, can I come looking for your soul!?"

Orthomolecular Therapy. Yes. Maybe some of that is what I need. Ortho — perpendicular. Growing out of. Coming out at right angles to what is usually the case. Indicates another dimension. A cube coming out of another cube. Tesseract.

Yes. Literature is a self-produced orthomolecular therapy for the management of split selves.

The Orthomolecular Chemistry of Hope

Two On Pillows Dreaming

We are watching two people in a bed. Laura on the left side, Walker on the right side, each within their split screen.

Left hers...

Fiery red hair billowing on a pillow. We see close up of youthful girl face; her eyes closed in sleep.

Right his...

Brown long hair, curls, angelic, but not like the girl's. We see close up of his long angular male face, closed eye lids.

View pulls back a little, to see both her and him in the frame. We see light and shadow dance over her face.

A shadow in the room?

He is in a darker, shadowy edge of the bed; shadows play across his face too and the room.

Hear: Wind blowing outside the windows of the room.

Camera pulls back in to see whole bed — we slowly zoom / pan over the hilly terrain of bead clothes billowing like landscape. Shadows moving on sheet suggest movie screen.

Camera pulls way back from bed to see the play of trees outside the window dancing in a windy moonlit night. Tree movement outside shivers and this projects shadow movements from the moon beaming into the room. Seen from below the trees are looming, looking in.

Camera zooms into his face. It fills the screen.

Screen fades to black then into the scene of his dream. On the screen now we see the scene of what he is dreaming. He is walking through a moonlit garden. It is the rose garden in Zilcher Park, Austin. It is after hours on the moonlit night.

We come to a pool; it is Deep Eddy pool. He enters the water and is swimming the quiet breast stroke.

We hear water; it is rushing in, a falls or a flow.

The screen of his scene fades to black.

The screen of her side fades in. We see her in her screen on the left side. We see her head on the pillow.

Camera pulls back to see her delicate curved hand in a warm light, it is curved like the fiddle-head of a violin.

She stretches out a bare foot from beneath the covers.

In the view of the bed world we see the two people. He pulls the covers up snug. And moves closer to her to feel the warmth of her body.

Camera zooms into her face. Her sleepy visage fills the screen. Screen fades to black then into the scene of her dream. On the screen now we see the scene of what she is dreaming. It is in black and white. She is walking on a narrow walkway across the slope of a hill. Below we see some roof tops of houses. She comes to a small store selling religious equipment; there are rosaries hanging in a window. Laura is walking down outside stairs cut into the side of the hill to get to this place.

There is a priest and some church ladies. They are looking at a framed photo of The Virgin Mary. The glass had been broken. They are upset. They shake their heads in dismay as they walk away. Laura stays there and starts to pray the rosary in front of the picture.

Close up of alabaster statue of the Blessed Virgin Mary's serene face. Camera lingers on the face, slowly panning over the bridge of the aquiline nose of the Mother of God. Camera centers on the closed eyes. The eyes suddenly open, they look out at us. The statue of the BVM levels her gaze out at the viewer. Her lips tremble as if she is trying to say something. But no sound comes out.

We hear Laura's voice-over while looking at the image of the virgin. "I take a pregnancy test and I am looking at the indicator in the light of the bathroom window. It turns

positive. I look with dismay into the bowl, at the water almost still. But then I realized: But wait I'm a virgin! So I am unhappy in the dream because I don't want to be pregnant.

"I consider an abortion. Next I am with Wendy. At the Planned Parenthood clinic. She asks me, 'Who is the father.' And I tell her, 'Walker.'"

"I was in the waiting room of the abortion store when I saw this little girl sitting on a chair crying. I went over to her and it was my little sister, she had a white dress with a little toy bear. She was a very pretty little girl but she had sadness in her face. Like she knew I was going away and wouldn't be able to play with her any more.

"But then next I am telling the nurse, 'I am a pregnant virgin!' And it doesn't seem to phase them, like they are winking and laughing behind my back and they are saying 'Yeah, right. She's a virgin.' And I get agitated. And rush out of the clinic."

Close up of her on her side. Her eyelids are twitching in rapid eye movement. Her frame fades to black.

His frame fades in. We see his face, then inside his dream.
On the screen we see in his dream as he is walking in the moonlight. It is an ancient stone building, perhaps once a Mayan ziggurat or a labyrinth, now open and overrun, with roofs gone. In the scene, shadows of wind blown trees are playing a chiascuro across his torso as he walks among ruins.

It is a black and white movie from the time before color, long ago. He sees his mother's face as a young woman. It is in an old black and white photograph of her childhood.

The photos seem to be floating on dark water, as though we are looking down a well.

In black and white glimmer of water undulations the photograph image is floating, gently rocking.

The photographs show a woman with an alabaster milky complexion.

Slowly his frame fades dark.

The screen of her side fades in. We see her in her screen on the left side. We see her head on the pillow.

Camera zooms in on her head. Screen fades to her dream.

On the screen we see Laura in the hall of school. She is running to class because she is late. She is holding big books to her breast. She is crying.

We see her boyfriend Walker as the school teacher.

In her dream, Mr. Underwood is in white shirt and tie, standing beside some lockers in between doors to classrooms. It is dark and the hall of lockers extends forever, into a soft light. Everyone else is in class.

He asks, "What happened?"

Laura wraps her arms around him, (books are conveniently gone). She blurts out: "I can't hold back how much I like you."

Then she steps back and looks him in the eye. She asks him straight out: "Do you like me too?"

And he says: "You know I do. And I do!"

They hold each other tight and kiss in the hall. He kisses her paternally on the head and says again, "I like you so much."

Her scene fades dark.

Camera cuts to scene of the couple in bed. We see her face smiling in the moon shadows.

Camera pulls back slowly to scene of couple in bed.

She ooches her backside toward him in the bed. He snuggles up closer toward her, draping his arm over her. We see two heads on the same pillow.

Camera zooms floating in and we see the two sleepy heads. They are close together in a shadowy, blue-moon room.

It is so shadowy, and moon-imbued, that the faces turn a

kind of blue as the shadows play over them.
Camera zooms in on his head. Frame fades to black, then opens on the scene of his dream.

We see a face of Krishna.
It is a head with eyes sweeping sideways, like those Indian dancing girls.
Krishna is wearing a tall hat that is stacked in concentric layers like a temple hat.
Camera moves in close on the hat.
The temple hat has all kinds of little charms, like from a charm bracelet hanging from it. This great magic temple hat has bits of golden microchip pieces inlaid, and little antennae coming out of it, to pick up transmissions. Krishna is listening to some radio program, his eyes looking up.
Camera pulls back and we see that Krishna is standing on some Mayan ziggurat, there are other ruins of temples around and big fields and the forest beyond.
There is a jaguar moving through a dark hall doorway of the ancient abandoned temple ruin.
Camera moves in on the prowling cat's sinewy musculature moving. The spots are moving. Floating in a saturated darkness like the water glimmer in the well.
Erie sounding winds blowing through the temple buildings, are moving the trees nearby.
Camera pulls back from the moving trees and we are in the bedroom looking across the couple in bed, out the bedroom window as trees outside move in the moonlight.
Scene of couple in bed:
They are both lying close together on their sides, she is curved into him, like spoons. Camera hovers and we see the two faces in peaceful slumber.

Night Movies through the Semi-permeable Reception Sets of Dream Radio

Home on the Range

Rainy days were good because I didn't have to go to my job at the construction site and could stay inside and write. Will Bill wouldn't have to be at his booth on the Drag, so he was home too.

To look at me and Wild Bill, we were pretty dissimilar. He was 40 and taciturn, sanguine; I was 30, overly solicitous of other's welfare, too eager to please — this from a basic insecurity. We were both shy men.

Wild Bill's physical appearance and presence was that of a Viking or a Greek god or a Sadus. He was a character. Out of an old time realistic Western. He had a big head of wild long hair that he clamped down with a leather head band or under a distinctive caballero's leather hat. He had a long salt-and-pepper beard that hung down his chest. He could have been one of the ZZ-Top graybeard bard archetypal Texans. He was at home living out of his truck and camping out in the woods. He could have passed for a prospector or an amateur archaeologist or a hippie searching for mind bending mushrooms and cactus. He had three dogs when I knew him in Austin. He was a local figure around Austin and quite recognizable: "You mean that guy with all the dogs?"

He was tanned dark from being out in the elements, and the dignified long beard made him look like a man you might see in an old, turn of the century, daguerreotype of men from then. John Muir comes to mind. With the headband and the ripped build and thick thatch of long wavy hair he looked like Samson the weight lifter from the black and white world of the 20s. Though he gave the impression of being a giant, he was in truth about 5'5 or 5'6 and very stocky.

He'd sit at the kitchen table and do his correspondence, tapping it out on his ancient portable typer or by hand on yellow legal pad. He got and wrote lots of letters. I was a good correspondent but he was epistolary, using letters like they did in the old days before phones.

We'd start the day drinking good coffee. He introduced me to the concept of drinking less of really good coffee. He'd brew a cup up in this little single cup coffee system that he had, that you had to grasp and invert to get the coffee to run through the grounds. It too was blackened from many campfires, the handle burned almost completely off. The coffee was delicious. I asked him what kind should I get over at the Whole Foods.

"Celebes," he said. I also bought a coffee maker like the one he had for myself too.

Well, now that Laura and I were in a relationship it was sweet to see her at the breakfast table too. She usually didn't stay overnight when she came over; I would walk her home through the sidewalk neighborhood at night. Occasionally she would spend the night. This required orchestrating an intricate story, weaving lies to her parents about sleep-overs and slumber parties and getting her girlfriends to cover for her. I'd make her a coffee in the morning and try to get her to eat some toast just so that she'd tarry a moment before she rushed out on her way to catch the bus to high school.

It was a delight to see this babe grace our old boy's table with her bright youthful presence. She might even take out a notebook or be arranging some papers from her back pack. Sitting there smiling, with her dress and her big cowboy boots. In a dressy sweater, cleverly packed to have a change of costume to throw mean nosy others off her business. Sweet Laura, running her hand through her hair.

Bill would come through, "Hey, morning."

She had good girlish energy, lithe and moving around, collecting her things.

"Thanks," she said, and headed out the door.

She tried to present as just an average shy teenager. It was a bit funny, here is this beautiful young woman trying to pass as a child. Though compared to her friends she was a bit on the thoughtful side, a bit serious, a bit sad, around these friends she was also wild and out there. She lives life as full as she can, knowing full well that this is the only chance she has at being young. Got to admire here gusto, if she feels like a swim, she jumps right in, clothes and all. Laura has a way of doing what she wants, letting nothing hold her back.

In the kitchen on a rainy day Wild Bill would start a new pot of his great chili beans. He'd stand up and stretch after his morning work out — he was religious about that workout — and announce: "Well, I guess I'll start my fire beans."

An excellent cook, he made some of the best chili beans on earth. He'd cook it up in his black 1940s-style pressure cooker with the valves and gauges on the top that you screwed down with butterfly clamps. He'd set it to simmering on the stove for hours over low heat. This pot too was black from many campfires and smelled of outdoors. It was great what the pressure cooker did for soy beans.

He'd start by pouring out the dried pintos and soy beans onto a plate to be sorted through by hand, removing any deformed ones, (or rocks) by dragging them out individually with a finger of his surprisingly slender hand.

"My father used to do this."

With his big hunting knife on a wooden hardwood board he'd chop the onions and garlic and jalapenos peppers.

Who boy! The chili beans that Wild Bill made. It was awesome, a hearty mixture of pinto and soy. He liked to use

steak or brisket that he cut into cubes.

He used fresh garlic — mashed and chopped; onion —sliced; and fresh jalapenos chopped, with their seeds left in.

"Don't clean the seeds out of the peppers cause that's where the heat is."

To eat Wild Bill's chili beans was a powerful culinary experience. After a couple of mouthfuls there started to be a round heat all over your mouth and it made you sweat, made you blush. Next the lips began to tingle and swell, then there would be the onset of a whole body blush leading to that heightened sensitivity of the senses and a bracing attention that was the reason why the gourmet of heat enjoyed a profound respect for the peppers and what they did for the savory of your being.

To somewhat mitigate the fire and make a more perfect protein, he made these delicious corn pones out of polenta, salt and boiling water. He shaped them my patting them between his hands until they were a good-sized, elongated, flattened mound, with finger indentations waving through the gritty yellow surface. These he fried to a golden crisp in oil — preferably bacon fat, that he kept in a big coffee can on the range between the rows of burners. These corn pones were like little islands in the great plate of the chili beans that were flowed all around it. The pone could hold their own for a while but slowly dissolve like castles in the sea, though they were not around long enough, for they got snarfed. This dish was incredibly good. He usually mashed his beans with a fork, being very methodical about it. I started doing that too. It is the proper way to eat chili beans.

I'd bake my bread which he liked.

Wild Bill was a fastidious, clean-freak, (Aquarius) house mate —quick to mop a floor or scrub a tub. He had asthma

and abhorred dust. He often gave his dogs baths so they lounged around our place with their glossy coats just inviting one to pet them. They were his children and they glowed, they beamed love at him.

He had developed his lung capacity, through jogging, and lunges. This also developed his quads, so that no matter how loose cut his jeans, they fit him tight as a second skin, over massive quads and thighs. His torso was as long if not longer than his legs, his shoulders were massive and powerful and made a long rounded slope, into lats that were so big from flys they looked like wings. He was rippling and sinewy, and graceful like a horse and girls couldn't keep their eyes off him. He could fight anybody, could be sudden and powerful and fast. This peripheral attentiveness of reach and grasp is reflected in the style of his writing which was sinuous and muscular, full of rushy speedy grace.

I'll get into matters of style more later, but for the moment let me just sum it up here. Through his writing, Wild Bill works to subvert the traditional set of moral values of readers of westerns. In *Tales of the Texas Gang* he celebrates a beauty of rhythm and flow, explores the mythology of his own personal life and relationships, raises violence to an action art, and enjoys the revelations of a transcendental connection with nature. Like an action painter he uses syntactical line and form to encode a deep sense of moving animistic destiny.

This is how he talked about the Beats:

"I dug the Beats in a notion male poetry can be a shout, after reading LEAVES OF GRASS and ON THE ROAD, back when falseness of demeanor carried more etiquette than it does even now in the U.S., back in white working class Texas in the fifties, late fifties was it. Thus, gushy Kerouac did arouse many such of us manly louts, from that drinking class. Most people may not remember, as I do so well, how most men were not so secure in their image, to be verbalizing

on a full moon, or past a limited sentence say. Or on trees or flowers or birds or Gulf waves that change every day much, and to forget allowing anything get a little further out. Before subtleties of the intellectual Beats had ignited the less intellectual hippy revolution, most young males had not been about to stare off spaced out, as had I always, who could fight."

Wild Bill was a true Texan, born in Aransas Pass. His father had worked in the refineries around the Gulf. His family owned property near Medina Lake outside San Antonio. I was an immigrant, born in Montreal but grew up in San Antonio. I always felt like a stranger in a strange land, my friends called me an iceback and I felt more affinity with the Mexican illegal immigrants than I did for the whites I grew up with. We didn't own any property.

We both had a commitment to writing. Wild Bill was a published author and had a book, that was to me a masterpiece of humor and style and most of all liberation. I looked up to him as a literary father figure. I was not only learning how to write the Real, I was learning how to live on the fringe as an artist.

Wild Bill dreamed of returning to an earlier time on the frontier, where people could be more natural. I was inspired in this too. On our walks we were always conversing about many topics. Often we did drugs: mushrooms, peyote. On the dog walks we became peripatetic Platonists of old, slowly learning to recognize the shadows for what they were and to turn around and look at the fire, the source of being from which the shadows were projected.

We though of ourselves as hunkered down in a world run by the artificially fit. He had this to say about the artistic calling and its requisite negative capability. "Read MOBY DICK. Melville was a hippy. The whale is spaced, cosmic, not in smaller focus while he basks on the sea surface building his oxygen. His evolution did not include men in boats coming

to kill him . . . For be a man's intellectual superiority what it will, it can never assume the practical, available supremacy over other men, without the aid of some sort of external arts and entrenchments, always, in themselves, more or less paltry and base. This it is, that for ever keeps God's true princes of the Empire from the world's hustings; and leaves the highest honors that this air can give, to those men who become famous more through their infinite inferiority to the choice hidden handful of the Divine Inert, than through their undoubted superiority over the dead level of the mass. Such large virtue lurks in these small things when extreme political superstitions invest them, that in some royal instances even to idiot imbecility they have imparted potency . . ."

He was good company on the dog walks and it was always fun to go with him. Off we'd go, Wild Bill and Walker and the three dogs Sissy, Griz and Cayief, riding around in a beat up old grey 1965 Dodge truck, with broken down springs sticking through the driver's side seat, so he had one of those folding back rest seats covering it. Inside the cab there were rubber floor mats. It smelled of leaky oil and brake fluid and there were assorted dog bowls rolling around. We headed out to the river to walk the dogs.

His favorite topics were: Spirit, Nature, Dogs, Indians, Drugs, Women — not necessarily in that order. He might start a topic with "What do you know about the American Indians?" Wild Bill claimed he was unusual because he was raised among the Comanche.

In his book *Tales of the Texas Gang* he had developed a realistic, countercultural, revisionist sampling of the history of the American West by putting his character "Wild Bill" right out in the wild lands with the Indians and settlers. His character feels their spiritual connections with the world and their anger as perhaps no author has ever done so well.

He would often say he thought it was the greatest period on earth, when the Plains Indians had the horse and the repeating rifle. It must have been supernatural. All that freedom and natural wildness before the onslaught of the white man. The *Tales* were all about that time. It was a most realistic "noble savage" fantasy, the gunfighters and the code of the West. All the characters that made up the Texas Gang were real. They were friends he had grown up with, most of whom were Viet Nam war vets who had been trained to kill and then set loose in that rock and roll slaughter house. Then one day just flown back out of the jungle to a street in the US amid hippie antiwar demonstrators, who were down right rude with their political correctness. He gives voice to his friends who were haunted by the horrors of the violence they had been asked to perpetrate in the name of America against ancient peoples. These blood brothers, *compadres*, were comrades at arms, but in Wild Bill's story they were natural men defending their way against an encroaching domestication that was stultifying. Wild Bill jokingly told me he would have to have a lobotomy in order to hold down the kinds of jobs he was likely to be offered.

For Wild Bill the Tales weren't exactly a fantasy. It was what he believed was a past-lives memory. Occasionally I would see him in this. We'd be walking in the woods, and he would have a transmigrational memory of a former life, he would feel the itchy trigger finger of the gunfighter that he was in a former lifetime. Or he might suddenly go into a muscle memory of the glide away from the thrust of a knife. He wrote from seeing a dream-self in another time and place. You have got to admit, as a student of writing, this would be the best possible ability for a writer to have. It definitely amplified for the reader, that sense of being present in what he saw through the shared vision of writing.

I wanted to be of service to this literature, which I thought was important. I wrote a review of *Tales of the Texas Gang*.

Compared it with Cormac McCarthy's *Blood Meridian*, which I couldn't get through at all because I found it completely humorless and relentlessly over-polished. I also compared it with *The Wild Bunch,* the movie of Sam Peckinpaw.

To me the prose flow of *Tales of the Texas Gang* was a great cross between the high speed spontaneous sketch style of Kerouac and the telegraphic beat style of Ginsberg, crossed with the gravitas in the imagery and storytelling of the Bible, and the Great Spirit transcendentalism of Melville, but with a lot more humor and a lot more fully developed countercultural authority. There was little or no wasted babble in the tales, just this fine old voice telling the truth. It was lean like Hemmingway and flowed fast and romantic and spiritual like Kerouac and had the moral authority of the Bible, while it hectically savored the teachings of drugs and violence and sex, and Mexican food, not necessarily in that order.

I made a note about *Tales* and the Bible. In those days one still occasionally would meet old-time Texans who grew up on an isolated farm and whose only contact with literature was the Bible. This influences their diction. There is some of this in Abraham Lincoln. Of course Wild Bill went to school but he did somehow, and I never got the story really, he did spend some time inside institutions, a hospital psych ward and the Navy and jail, so it might have been there that he did somehow annex that phraseology. Maybe he heard it in his father's speech or other elder family members. The boys in the Texas Gang too also spoke in Biblical phraseology. Most obviously, they peppered their speech with *Beholds*! and *Foresooths*! at revelatory experiences. This lent a great level of craft and verisimilitude to the work.

"What do you know about the American Indians?" had been the question.

"Jeese, I don't know much, just what everybody knows from seeing them on TV. Tonto talking to the Qui Mo-Sabe.

What the heck does that mean, anyway? Some kind of Tex Mex / Indian patois. Qui Mo-Sabe."

"The one who knows more?"

"You have to know something if you are going to know "mo" of that something."

WB: "We played cowboys and Indians when I was a kid, and I always wanted to be the Indian."

"Me too."

WB: "On TV the Cowboys were the (Good Guys) and the Indians were the Bad Guys, always attacking the circled wagons and always portrayed as threatening to massacre women and children and such. So stupid.

"If you got any knowledge at all about the aboriginal humans who came here to live first, it was from the "standard text" adopted by the curriculum committee and allowed by the Fed and the State boards. You might have heard of one or two famous Indians, Sitting Bull and Crazy Horse, maybe some high profile battles."

Wild Bill spoke with contempt of the U.S. Cavalry and Little Big Horn. "They wouldn't have told you about how 'ole' Yellow Hair' General George Armstrong Custer was the designated director of exterminating the 'savages' or 'red man.'"

The first people were defending their world, trying to keep the waves and waves of immigrant hoards from hogging all the resources and draining their livelihood and way of life and culture from existence. You only hear about the struggles of the pioneers and settlers who made the great trek and went from fort to fort. The western movies portrayed the Indian as just this evil lurking out there in the environment.

It was from Wild Bill that I really learned about the true, the original indigenous Americans, who braved the times of this vast north American continent. No 7-11s, no Howard Johnsons, no motels with air conditioning and TV's, no

electricity, no hospitals, nor beauty parlors. They were basically poor financially, could not afford a BMW or monster truck. But they were rich in psychic culture and spirituality.

    Wild Bill would go off, about these native Americans / Indians who occupied this land hundreds of years before Columbus, the Pilgrims, and all the white trash immigrants crawling all over each other like a bunch of insane insects trying to own everything they saw. These mindless white men who felt like they had to be always in charge, carried out extermination of the Indian, or forced them to live on reservations in areas that were devoid of game or livelihood, so that they became dependent on the white man military for provisions necessary for survival. Unemployment. The lazy Indian. They just did not want to work in the white man's mine, or lumbering their beloved forests, or rendering out their hormone fattened cattle into hamburgers. The US Cavalry and government used germ warfare: issued these freezing, defeated peoples blankets which were INTENTIONALLY infected with small pox, that was fatal to them. The We the People of the U.S. Constitution did not include the Native Americans. To the whites they weren't even people. I was starting to realize that those early Americans framers of democracy, (transplanted European immigrant *outremonters*) saw that these indigenous peoples were too different from them — in their religions, their language, their way of being in the world — to have been co-opted and assimilated. So it was decided to do genocide and ethnic cleansing on them until they absolutely ceased all military activity.

    Wild Bill had this biblical style — by that I mean a sense of divine destiny, a sense of the rapture awaiting beyond the edge of life, a sense of moral outrage and righteousness in the rhetoric of his images and action — that allowed him to

map these sensitivities into his own life and reflect modern everyday concerns in his art. Wildness is an essential part of our lives and to loose it is to loose our souls.

He was sure of himself, and he prided himself in not carrying guilt. He was relaxed and he helped those around him to be relaxed. He wrote from instinct, from intuition. He was in touch with the spirit.

Well, Laura and I were doing a lot for each other.

I was enjoying it. I was definitely feeling better. I was in love or just happy for once. And I had a lot of interest and patience for everyone. It was the last Spring of my 30th year, a year in which I had suffered a most egregious breach of my security when the cops busted my little marijuana farm in Manor, outside Austin and I had started life all over once again after getting out of jail with nothing but the few clothes, and scattered manuscripts and a couple of dollars.

And now here was this lovely young woman come into my life. Her thick auburn tresses cascading over her shoulders in rich silken waves fell down on each side of her eyes, almost shadowing the eyes. She had freckles and a pouty mouth. Her eyebrows were a bit dark. I know she didn't use any make up. She could have this almost shy provocative snooty way of looking at you. Like some wild thing, that had graced your presence. She could look a little sad and tragic, carrying too much responsibility for one so young. She was a tall strong woman, her glorious hair fell down passed her regal neck and shoulders to cover her breasts. She moved relaxed, elegant. That's what we liked, she felt OK around here, in charge. It was a gift to give the teeny this power over me. I wonder if she knew. Yes I think so, the way she held my gaze. One day she had on a pleated tartan school-girl skirt. She walked in front of the light in her little top and short skirt. She rolled her hips and stuck her bum out and

winked at me, enjoying her power. She wore a little ribbon around her neck sometimes, it was like a cameo. It made her look so prim, it was delightful. Though she might have been putting on an act, more sophisticated than she was she carried if off very well

Laura was sweet when she came over. And we were getting much better at doing it. And so, since we were lovers and had thus become Ambassadors of Love in the world, I asked Laura if she might know someone who might like Wild Bill. I said it would be nice to fix up my friend.

She brought her friend Wendy over and Wendy and Wild Bill hit it off. Yes. Wild Bill —squaw-man, dog-man, horseman, man's man liked big warrior women, and Wendy was exactly that. Pretty soon they too were out doing their relationship.

It was just the sweetest thing in the world for the four of us, two old bohos to have these two exceptionally sharp and independent young high-school girl / women coming around. How did we deserve to be so blessed? The two young women started coming around more often. After the dogs and people around Baylor Street started to know them, they became like regulars. There were so many dogs around we didn't need to lock the door.

They would show up with marijuana and beer, and sometimes other drugs. Who knew from high school girls? They fly all around get into scenes, score the best smoke, fuck anybody they want to, and then be home in time for mom's spaghetti dinner. Well, Wendy didn't have a mom, she was more like a kind of mom. She was mature and used to being in charge. She was quite the favorite of her father's gay friends. Wendy of the lost boys. More than once I would come home and find the two of them, Wendy and Laura sunbathing topless on my back deck. People just started going tribal around Wild Bill, it was an effect he had.

Guarding Angels

In addition to their other responsibilities, the teenage girls were involved in baby sitting. That's how they made their spending cash. Laura sometimes had her little sister Elizabeth with her. Other times she brought a girlfriend from high school, Isabelle, who took care of some young boy, her brother Jesse.

Isabelle, or as she liked to be called — Isa, was as I would come to find out, quite a piece of work. Though she might be dressed demurely in a serious school-girl uniform there was an amazing punk energy surging beneath the facade of student/girl/nanny she was required to portray.

Sometimes when Laura and Isabelle were bringing their young wards to nearby Pease Park to explore Shoal Creek or the playgrounds nearby, they stopped by for a visit at Baylor Street. It was a cool safe place, with tree houses even. And dogs to pet. The tree house was easy to get up into if you went on Talbot's porch, up a ladder through the roof, and across the walkway. I would take them up there; keep an eye on them. Luckily no little people took a noser out of the giant oak. It was pretty safe.

Something happened when these girls did baby sitting, a beautiful side of them came out. It was very childlike and resourceful. I felt like an anthropologist observing human behavior, somewhat removed, but also privileged. It is the writer's blessing / curse to be in the scene, but also watching the scene.

All the time that young females spent talking amongst children and telling stories and songs as guardian angels gave them major verbal abilities, not to mention sweetness skills. Neurologists maintain that women have a fiber-optic

cable running from the speech centers of the brain to the mouth. Teens are all working on their identity, but the girls could style out among their girlfriends about love and goddess and power. This was something that ordinarily guys are not permitted to see.

Isabelle really liked doing the show for Laura and me.

Walker, Laura and Isabelle are walking in Pease Park. Toward the oak trees along Shoal Creek. Laura and Isabelle have guitars. It is near the children's playground in Pease Park, a couple of blocks down Baylor St.

Isabelle is walking, and talking to Laura as if I wasn't even there, "We gotta beware of the communists, or else we are all going to have to learn Chinese."

I followed along, the squire of one and the dad of the other. Isabelle said, "God how I hate school."

Then Isabelle shifted into some kind of humorous rap about the strategy of child molesters.

"Well, like. . . OK. The child molesters. They have like dogs?" . . .

She raises her hands together acting as a somnambulist being drawn toward some irresistible attraction. She moves the raised hands toward each other, illustrating attraction. The hands cup around a precious ball.

" . . .And children just swarm in there." . . .

Hands floating, gesture as if she has got a hold of some awful swarming ball of copulating garter snakes. Isa feigns a look of mock horror. She gasps, as if recoiling from the disgusting scene.

Her dexterous girl-fingers are gnarled into knots, she is beholding / acting — portraying the lascivious predator ready to pounce and clutch. Pease Park is green in the background.

. . . "Can I pet the puppy?" . . .

She folds her hands together in beseeching prayer as she

enacts the pleading imploring gesture of a child.

She spreads her hands in a pontificating gesture:

. . . "And they're like *Duh*. I'll take pictures of you while you're bending down. And I am looking down your shirt."

She points to her own flat girl chest, tossing her lovely long hair. She is teasing and being seductive while at the same time, invoking your protection and admiration. Like how the prim, cotton, school-girl uniform contrasts with her loose gestures and go against the grain of what she is saying. . . "They're like *awwwe* puppy. What's his name?" . .

Isa looks aside at you and bats her innocent eyelashes. She has olive skin, and naturally dark shadowed eyes.

She hits you with her great come-hither, get-away block, to which even her mother is not wise. And you see that she is a powerful phenomenon that promises the most intelligent intense love. She is dangerous and beautiful.

. . . "And it starts a conversation — leading to more.". .

Isa looks at her audience all blasé, glamorous — like she has seen it all before. Walking through Pease Park, it is her way of styling. She gives us a feigned helpless look, with an enticing disregard for consequences or reason as if to say: I can't help it. / Everywhere I walk on the street at the mall / I see other men getting ready to pounce on me. / It's not my fault.

She is Punk, trying to be outrageous but not too far out of context. She can't suppress a bust-out grin of smirking smartness as she says, "And, you know, they get it on."

(I wondered if she knew what she was talking about.)

She raises her eyes, as if to say she is imagining herself being pursued by a pack of wolves.

She sticks out her tongue, rolls it around as if to taste the air for the residue of her cuteness, and sensing she has captured us with her play, she smiles rueful.

... "And she's like
(Shifts into a high child voice)
... "— Wait! I'm only 7!" ...
(Shifts back into her own Isa voice.)
... "He's like ..." ...
(Shifts into a deep male voice.)
... "I'm 49." ...
(Goes into narrator voice.)
... "She's like
(Shifts into a high child voice)
... "— Oh we can't make this dance work." ...
Isa starts to laugh.
She turns away, in a gesture that means Oh the shame and horror of it all. She is also grinning and enjoying her story telling ability.

Another time in the park.
Isabelle singing *Jesse Come Home*, with all the slow feeling and plaintive longing that Joan Baez or Janis Ian or Roberta Flack puts into it.
Jesse, come home / There's a hole in the bed / Where we slept. Her little brother is writhing almost epileptic in the throws of embarrassment.

Another time in the park, Isabelle sings like she is on stage. Isabelle is taller than Laura, and rangy, very beautiful, this nubile punk rocker girl. She was wearing very young clothes, candy-stripe knee socks over panty hose under a short skirt and a crisp top. Charm bracelets clinked on her wrists as she gestured dramatically; she kept looking out at me from under a head of thatched hair — a young smart girl with a hopeful face framed by pendants hanging from her ear lobes. Slender to skinny, she played basketball though

she was only a freshman. She seemed way older than her 14 years, although she looked to be about 11, the way she was dressed.

Isabelle sings fast and clipped.

"Talk you talk the talk all talk"

Like a machine gun "talk-at-ta, talk-at-ta, talk-at-ta" she is accentuating every syllable.

She is dancing some two step, but fast, like it was the crossover in basketball. The nubile girl, tall and wild, gangly as a Watusi, born to play basketball — her hair lofting when she tosses her head as she swung around suddenly, dancing some fast step.

(She shifts into her low dusky voice.)

"This boy is never gonna be mine, / this boy is going to be someone's / but he's never gonna be mine.

Things are going somewhere some time. . .

(Shifts into a high keening voice :

"Heee walks along / Heee walks along . . ."

(She slaps her hands on her thighs emphatically, like someone grieving — terribly, insistently.)

"It's my nnhhhaa"

(She makes a sad face at us, but winking too. Shouts:)

"When you smile, Yeah!"

(She jumps up in the air flapping her arms.

"When you smile, yeah, yeah!"

We are standing there, I am kind of shocked, Laura is laughing and the laughter is egging Isabelle on.

Now she has her back to us, her audience and bends way over and shows her butt to us and looks up at us from between her coltish legs, the short dress barely covering her small behind atop the white leggings and striped pink knee socks. She looks up at us from between her legs, her hair hanging down. Then she jumps up and turns around puts her face up, mugging, jumps away twisting back and forth. She

was gyrating around, totally nuts. She sings:

"Uh, we are meeting at the mall / we are walking side by side in that silly slide / where everybody is / going to shop / shop till you drop, / that's all they want to do /buy up all the stuff and take it home.

(Shifts into a sophisticated southern blues singer:)

How did we become / this white bourgeois / so addicted and so lazy / what a bunch / of whiners at the free lunch.

I won't let it happen to me / I won't let it happen to me.

It's not the path I chose to take.

It's not the path I choose to take.

Another time in the park. She is walking.

Isabelle started singing what might be her theme song.

(Sung to a rocking walking beat, a "we are following some pretty girl down the sidewalk from behind" beat.)

Voice Over:

I am Isa

My parents are hippies / migrant mestizos / wetback dope dealers.

Some call me Whessa

I am Huisa

A storm, a fire, that all desire.

A phenomenon, a common con,

Going to be the head of the class

I am Whessa and you better not be looking at my ass.

Another time in the flat, Isabelle and Laura were disco dancing. I found myself quite taken with Wheesa's wild child ways. It allowed Laura to appear as the more sophisticated, reserved caretaker one of that pair. I liked them both. Isabelle is a woman too, though she is only 14, a very precocious 14. So close to the wild heart of life. Her little tits were escaping freely out of her halter top.

Agent Felicity

Wouldn't you know that Isabelle's mom turns out to be an old friend of Wild Bill's. They go way back, for Isabelle's mom was his dope dealer. Theirs was a long association; she had dated one of his friends in the Texas Gang. She was a good old Texas girl with a relaxed knowing smile, from whom Wild Bill occasionally bought marijuana and other substances, when available.

I had occasion to visit Isabelle's mother once or twice, in the company of Wild Bill. Of course we did not make any mention of her daughter's behavior — exercising her considerable powers of lolita hotness at our place under the appreciative and respectful eye of these two good old boys. The mother's name was Claire, we called her Claire — never knew her last name. Don't recall her last name. She dealt a great artistic product and people were much helped and encouraged and comforted and made oblivious to their pain and myriad shortcomings by it. I do know that she had been in the marijuana purveyance business for a long time and you don't do that unless you are extremely savvy, streetwise, careful and hip. Over years of being in the illegal trade she had developed the dope dealer's felicity.

Forgive me if I digress for a moment on a beloved character of my youth. A good dope dealer is a kind of urban shaman who has emerged as a facilitator of consciousness. They are street wise and savvy, cautious and aware of what is going down around them. Also they have highly intuitive perception into what people are feeling, what they need. Being a good dope dealer is like being a fine chef preparing a substantial feast, a Eucharistic feast for savoring sensors

in the mind as well as the body. Are dope dealers the agents of the herb? Propagators of memes? Yes. They are doing the good work of our savior — Sister Mary Jane.

The whole context surrounding weed is a big part of the trip. Let us not forget: if you smoke you deal and if you deal you're an outlaw. This gave an edgy, shadowy context to the pursuit of recreational and empathic hallucinogens: the consequences of illegality were a very important element of tribesmen culture. So forgive me if I go briefly into this body of knowledge and manners around the experience of using marijuana for it is of anthropological significance. Though being an agent of consciousness might appear as new career path, it is performing the shaman function for a community. Who knows what kinds of knowledge future dope dealers might be called upon to know and become the propagating repository of. Perhaps they will have to have advanced degrees in Anesthesiology and Quantum Consciousness to guide the pilgrim in the stochastic traversal of possible ontologies. Perhaps they will be able to prescribe and deliver pinpoint dopants to numb specific maladies either physical or mental. Or their services might be integral for the exploration of parallel realities only accessible through perceptronic stimulants like some future *mathmethmeme*, say.

As I said, Claire was Wild Bill's dealer and not mine, so I did not know her, though we went in together and bought some weed from her. I had my own dealer of a long time standing, Kenny. Crazy Kenny we used to call him. We called him that because he didn't have a typical male response. He was supportive, kind. Interested in the spirit. It didn't enter our mind that he might be gay. He just seemed to have an even stronger Peter Pan Complex than most. He was a hippie from way back, from the anti-war movement, from the affinity with Indians movement. For the longest time he lived in a garage that gave out onto an alley. Kenny wasn't

in it for the money. He lived in a converted garage, he didn't have a car, nor a phone. He had it all nice and cozy in there like a wigwam or the inside of a teepee: small wood-burning stove, carpets all over the floors and walls — like animal fur. They were nice carpets too, with swirling, curlicue movement of natural fibers, like from Afghanistan or Turkey, one of them places. Inside, it was like an Arabian sheik's tent. It was an oasis, in the madness. There were great mounds of pillows. There were eagle feathers hanging from dream-catchers. The walls were festooned with rock concert posters. Later Kenny became the drummer in a punk band that played at Raoul's, a raucous nightclub in Austin.

We were friends and I crashed on his floor more than one night.

It was amazing that he was able to carry on his business without a phone. He didn't get a phone until much later. People just showed up. It was like a floating laid back party, not loud, though he usually had music playing. He had electricity back there, and a boom box. Lots of tapes.

I picked up the viper's argot gradually by assimilation. No doubt some of it and the folk technology of drugs Kenny passed on to me. Perhaps he was an exception in the world of dope dealers; I haven't known that many weed merchants in my life.

He wanted people to feel better around him. In his way he felt like he was part of something, a change of consciousness. Though he would never have put it in these terms, he was making the software to change consciousness available to minds.

There was a whole dimension of being and feeling out there that we could have access to with a little help from our dear gentle sister Mary Jane. The synchronistic as Father Jung called it, as Father Wolfgang Pauli called it. Kenny was doing work helping the divine helper, the generosity's work, and I

do mean big-G (God to some.)

In general we called marijuana *weed, smoke, boo-jay, MJ, smoking mixture, kind,* and *da kind*. We learned the terms for gradations of marijuana quality which were *working smoke* for ordinary everyday weed and *top drawer, top shelf*, or *personal stash* for the finer grades.

In the early days we bought a kilo brick of Mexican *mota* that had been crushed together in a kilo press. It came wrapped in Spanish newspaper. This weed was all pretty much an indistinguishable, sandy brown mass, with lots of seeds and stems. Acapulco Gold was good and only $100 / kilo. In those days it was important to clean your dope, to get the seeds out so that they didn't pop — causing the joint to explode in your face or to burn unevenly down one side — that was certainly the mark of a neophyte. Joints don't have additives and fire accelerants in them like the tobacco companies put in cigarettes to make them burn smoothly. The *mota* was very nice. Although by today's standards for the care and propagation of botanical beauty, it would be considered quite inferior. I wish I could get my hands on some of the Mexican brown today. Later I got exposed to some great stuff over the years, Colombian, Hawaiian. Thai. Even ancient *ganja* from India showed up once.

When the US government starting insisting on the use of paraquat in the eradication of Mexican marijuana fields, and the paraquat tainted weed started showing up in our baggies, I like many others started to grow my own. Paraquat did more for domestic marijuana production than anyone would have imagined. The continued high value because of prohibition also makes it the cash crop of choice in many parts of the country and the world. I became a good grower and was proud to show some great weed that was not only physically beautiful but psychobotanically beautiful in its high.

You can get many different types of bud. Mostly it is

grades: OK, good and great. The terms *Shwag* and *Nug* are used. *Shwag* is what we used to call *dog weed*,— leafy, dried too much. *Nug* comes from nuggets of manicured buds and came into use much later with domestic growing in California. We might occasionally see a Thai stick, a trimmed bud. The color of your bud usually plays a big part in how potent it is, and how high you will get. Usually the greener the bud, the more potent it is. Brown bud will usually be older and less potent. However Hawaiian for example, is lighter green and is very good. With more experience you will be able to tell what is going to be good and what is bad. The feel of the weed will also tell you of the quality. If the weed is soft and sticky to the touch, it will be pretty good. You will usually hear people refer to this as being *dank* or even *tacky*. It is the resin that the female flower uses to attract and attach the pollen to grow into a seed that is sticky. Crystals of this resin contain the psychoactive THC. It is these crystals of resin that the hashishins shake off the flower by beating and then collect and wring through cheesecloth into lumps and bricks that makes the concentrated goodness of hash used to assassinate mind and senses. If you leave your bud in the heat or a humid place, it will dry out a lot faster and become less potent. So try to keep your bud in a sealed container. *Shwag* is the most common and will most likely be what you are smoking. It usually comes with a moderate amount of seeds in it depending on the quality of the plant.

    It wasn't until later when I became a grower that I could distinguish between male and female marijuana plants, and really learned how to properly handle it and got some experience with the different strains. I must mention that the push now is to grow multiple crops of *indica* hybrid, which looks good and has a powerful odor — *skunk* it is called, which has come to be associated with *Nug*. The skunk weed

hybrids grew out of a desire to have a shorter growing period in a commercial indoor hydroponic production context. Also the *indica* varietal has a downer narcotic buzz experienced as a dulling sleepy effect. This state was much sought after by the 80s downer generation, people who were into Mandrax and Quaaludes and listened to head banging bands called Anthrax. It is rare now to experience the great philosophical and sexy *sativa* with its ancient *ganja* pedigree and Vedic provenance — Alas.

Generally *Nug* or *KB (Kind Bud / Killer Bud)* is a more expensive bud. This is the next step up before you start getting into named plants. Some names are God's Gift, Sour Diesel, Hawaiian Snow, Amnesia Haze. I worked on one called Texas Twister. In general *Nug* is much more potent than *Shwag* and takes a lot less to achieve your high. Because it is more potent, it is generally harder to find and will cost more.

It was surprising how sometimes Kenny had to be patient with people who did not bring the right amount of money, and I think he even extended credit, which is not wise in the drug underworld. When you're purchasing bud from someone, you are doing something illegal. So don't get freaked out if your seller is paranoid. One thing that can piss dealers off is if you give them a bunch of small bills for the purchase. It is good etiquette to give them the smallest amount of bills as possible. They don't have nickel and dime bags any more, now it is $40 for a quarter oz. Give them two $20s.

Kenny usually had a supply of his weed wares on hand at all times, and he was free with it, for after people had a taste of his dope, they usually bought some. A vendor has to factor this shrinkage into the pricing of product.

Later when hash started turning up we had to learn the metric system. He went to considerable effort to get our little hand scales to be accurate. We found that there were a little over 3 grams in an American Lincoln penny. And since there

are about 28 grams in an ounce, a quarter ounce was 7 grams, about 3 pennies. We never insulted him by checking his weighing, and he was conspicuous about being careful to get his measure correct. Soon a digital scale turned up.

I never saw any unpleasantness in his dealing encounters. Though I understand there often is some. My experience with dope dealers may be somewhat atypical.

Later on Kenny got a phone. Then a primary injunction against using the terms for marijuana over the phone comes into play. Exercising healthy paranoia, we just assumed (correctly) that the phones were not secure. Dealers hate people who call them and mention bud on the phone. You want to be overly solicitous so as to not even so much as hint at the suggestion of putting your dope dealer into risk.

Which brings me to the main unspoken thing: If caught you NEVER SNITCH. No matter what. Being a toker automatically makes you an outlaw. You have to develop some standards, some ethics, some code. Rational hedonism is a fun hobby, and you take your own personal chance with whatever risks or repercussions. If you get pinched, you accept it, and take responsibility for your actions. You don't hand them over someone's name, you don't pull others into your own personal hole just because you are too cowardly to deal with it alone. Besides, what are they going to do to you over a little possession charge? Even if you are a respected physician or a lawyer or a politician and can loose your career. You don't get to continue in it at the cost of someone else's freedom, the dealer who got you what you asked for because you said you needed it.

Kenny never dealt with kids either. They can't keep secrets.

To sum up: the dope dealer relationship is all about safety from police. Even though marijuana never hurt anyone out of the millions of people who have used it, there are some

800,000+ incarcerated. They've had their lives turned upside-down every year because of the police having to enforce an outmoded prohibition. If you can meet your dealer without even thinking that you're doing something illegal, you know you and your dealer are operating safely. For sure, that means no hand-offs in the open. Police eyes are trained to notice this. You'll end up in hand-cuffs.

Most dealers have more than one strain available at a time. It's okay to find out what they have and even ask about it. Most dealers have no problem discussing marijuana. But, keep it short and sweet. It is quite an education; you need to find a common language to find out what they have. At some level, especially after I read the Masters and Houston book *Varieties of Psychedelic Experience*, and Timothy Leary, especially one based on the Book of the Dead, I was aware that the accepting framework created by the dope dealer had an impact on the experience of the user, especially psychedelics. In the early days we were trying to find out anything we could about it, and it was truly frightening and horrible to read the "psychomimetic" (similar to the experience of psychosis) episodes in the excruciating field notes of the early psychonauts in the old journals.

If the dope dealer said that the LSD was all right and clean, then you could rest assured that at least you wouldn't be poisoned on your trip. In that sense a good dope dealer was a poison taster at the kings banquet.

The second rule of Rational Hedonism and Recreational Drug use is, Never get stoned 2 days in a row. You have to maintain your casual user status, no matter how much the unconscious persuades that it is OK not to.

An interesting problem for the dope dealer and client, and the increasingly more discriminating connoisseur, is the aesthetics of description. The kind of descriptive language like that in enology has yet to emerge, though it may exist

in some Vedic literatures and cultures. It is just too soon and too underground here. Look how impoverished the language is: Instead of saying something like The smoke assails the mind's perceptivity by surrounding it with a pregnant nebula of possibility — the dealer might say, it's pretty *whack*.

Or instead of saying: It will give you a case of the dropsy, your eyelids will droop, you'll forget the endless dialect of disappointment and dissolve in a dream — they say you'll get *baked*.

Instead of saying: You will enter into the province of the particular, peopled by polite politicians and everything will seem funny sunny; you'll feel like that special bunny you used to have on your bed for a sleep toy — they might just say: *blotto*.

Instead of saying: It will leave you with a sense of psychological isolation — *you'll get paranoid*. It will slow time down, objects will become < *italicized* >. You'll feel like wooing *. (You might find yourself looking around in the field for lost footnotes to the psychological nuances.)

In the future one will inhale the vaporized oils and crystals without combustion smoke containing tars and carcinogens. Also joint and muscle pain relief will be delivered topically in emulsified lotions; stomach knots will be untangled in soothing syrups and drinks.

I think a lot of your dope dealers later in life went back to school and got degrees in psychological counseling and social work because they had that kind of tough-minded agape in their basic personality. Being a skyhighatrist is a good background for becoming a psychiatrist.

I know Isabelle's mom did go on and become a marriage and family counselor.

Kenny never changed.

The Night Auditor

It was the ten year anniversary of landing on the moon and I wanted to have a feeling about that great event. I wanted to think about what was going on with me then, and now. And also, underlying the impulse to memorialize the Man in the Moon, there was this sense of time repeating itself but shifted in a parallel relationship with Laura. For I was in my first serious love relationship then, as she is perhaps in her first one now. That was a turning point summer for me then in many ways and might be something like that for her too. Hopefully by understanding my experience, I can make her experience good. The story of that summer gets at a profound experience I had then where I broke from my youth, and worked out that break. That summer of 69 is a focal point of my life, one that shifts into many strata of my being and one to which I return.

July 20, 1969. Man was on the moon!

For a while there, the tremendous triumph resulted in people smiling with recognition toward each other in the streets. Suddenly we were all made aware: photos of the earth from space, let everyone see a magnificent blue white marble showing the sign that we are here. That image of the sustaining earth from the moon is the most archetypal image of my generation. There was an aura of greatness as people regarded each other. There was a universal joyous recognition at the momentousness of the occasion. I was with the millions all over the world who watched and listened as a steady, jubilant voice announced across a quarter million miles: "Houston, Tranquility Base here. The Eagle has landed."

Man had landed on his most happy archetype, the moon.

It was such an epochal occasion that everybody will remember where they were at that moment. I watched the voyage to the moon on a TV in the royal suite of the Chateau Frontenac hotel in Quebec City. The Chateau with its 618 sumptuous rooms, its turrets and its copper roof turned cupric green, is situated high on a bluff above the St. Lawrence river. The house detective — a French Canadian heavy, complete with brown double-breasted pinstripe suit, stingy-brim hat and rake-hell moustache (looking like someone out of a Mickey Spillane novel) let us and a few chambermaids and other staff into the unoccupied suite. I was working at the Chateau as a night auditor. I was infatuated with that old castle and the job gave me ample time to dwell in it, to explore it. Le Chateau Frontenac is to Quebec City what the Eiffel Tower is to Paris. It is the most photographed hotel in the world.

In those days I had this way of traveling: hospitality work. You could travel around and get jobs in hotels, where there was immediate cash from tips, and usually meals in a staff cafeteria or at the staff table in a restaurant. And often, especially in these CPR hotels, housing for the staff. I had worked at the Algonquin in St. Andrews, New Brunswick while in high school. Now I was in the Chateau Frontenac accounting department, where everybody was French — certainly those of us on the night shift. Mine was a bit of an odd job: keeping track of all the minutia of tourists' purchases in their coming and going through the many small departments in this huge world-class hotel.

It was all very exciting. I was living with a girlfriend for the first time. My lady's name was Colleen and she was apparently pregnant (having missed her period) and we were separately thinking about getting married. I was growing up fast, taking the first step toward adult sexuality from the self-absorption of adolescence. (I was a late bloomer by the

standards of the Love Generation.) She had introduced me to sex that year back at St. F.X. university in Antigonish, Nova Scotia and to mescaline that summer. Sex was the awakening of a powerful love consciousness and I felt a part of the generations of lovers in all the world. I was her impetuous Texan, her lustful young Canook. We drove into Quebec City from Antigonish after the end of the semester in my old '53 Dodge hemi with posi-traction rear end and good radio. I had picked this used car up in San Antonio for $100 and driven it all the way up to St. F.X. It slurked oil like a binge drinker.

Now at the Chateau, there was a carnival atmosphere of giddy excitement around the moon walk: the staff ditching work, going, up and down to the royal suite to watch as millions watched that cautious "Giant leap for mankind." And we all stood with Neil Armstrong on a new world.
We were now much closer to the moon; it was strange, uninhabited; no atmosphere; its gravity 1/4th earth's, it is a levity. Over eons, the moon's smooth luminous spherical surface had entered through the imagination along with its constant companionship with the sun, into the archetypal realm of platonic ideal. Indeed the geography of the moon reads like a roster of the most enlightened, arid, meditative men who have ever lived on the earth: the Leibnitz Rills, Crater Tycho, Newton Mountains, the Sea of Tranquility . . .

I had become friends with the manager's son and with this scion of the king as ally in the court, we totally had the run of the place. I would go there early before my shift sometimes just to get into wandering around the palace with him. We used to climb all over that castle, going out the windows of a chambermaid's room in the attic, and scaling ladders over the cupric copper and slate roof up into the very top of the flag-bearing turrets.

This elegant hotel, built like a medieval castle, rises majestically like in a dream out of the stone promontory called Cap Diamant on the Plains of Abraham overlooking from high above, the St. Lawrence river as it wends its way in a vast river valley toward the distant Atlantic. It was built by the Canadian Pacific Railway in the last century. Winston Churchill had held a big conference here during WWII. Alfred Hitchcock stayed here. For me being in this ancient city with its little cobblestone streets, and coming and going through the gates of the ancient wall of the citadel, amid its Renaissance and Middle Age architecture was like being in a true life castle of people's dreams. Disney without the kids, but with history, and a magnificent outdoor terrace upon which to be part of the world promenade. I had just turned 21 not a month before and was high on love and expatriate idealism. The manager's son was enthusiastic about it too. He was younger and he found in me a kindred spirit about the hotel and he wanted to impress me with his masterful knowledge of the potala.

At night in the hotel we roamed the hallways, bars, dining rooms and kitchen — I was the night auditor for all the departments even the switchboard. We ran the teletype and the computer. In those days I'd get stoned and imagine I was the Night Auditor of the Universe.

In the magnificent Chateau Frontenac kitchen we ate like kings. Every night it was delicacies like escargot, exotic cheeses or steak; fantastic French pastries or triple-thick rich chocolaty malts. This contrasted with the one-room, communal-kitchen, bathroom-down-the-hall situation where I was living with my first real "old lady." Actually she was my first sexual partner, and we were both pretty naive. But very real and loving and true and idealistic. We had a lot in common. Life was a long deep conversation carried on in bed, at meals, on walks in the park, at discos, and

around bars. She had just finished her Masters in English Lit, writing her thesis on Samuel Beckett as well as being an accomplished actress. She had played in *Tom Paine,* and *The Bachae*, when we were at college in Nova Scotia. I was working on my first novel, reading Proust and feeling like van Gogh, alive, wading in a stream of energy forms, part of the sun.

When I got off work I would walk home through the awakening Quebec streets, whistling, smiling at the old women up early sweeping off the front porch and their part of the side walk. I was so excited to get back to the side of my love. I'd start to run the last few blocks to our room. I'd bound up the back steps, thrust the key into the lock and make a diving leap — living deep for her, dragging her jeans off as we tumbled lazily and gently into the bed. I was already ready and would play and tease her. First thing I'd do is get her into a hug then fall across the big brass bed, and hold her close. I loved that bonnie lass. And we had our own love pad in Quebec.

The next day, statements about the moon landing in the newspapers and on television from all the world's heads were circulating. I still like Trudeau's best: "Man has reached out and poked a finger in the tranquil moon."

Pictures came back to us from the moon of the beautiful blue-eyed baby earth floating in the blackness of space like an embryo curled up and swaddled all around in a placenta of clouds in a blue atmosphere. Man realized that we were on an island in the universe, and more than ever, were brothers and sisters. In addition to understanding the fragile dependency on the planet, man was on the threshold of leaving off his childish self-absorption and taking the first tottering step toward the stars.

The Night Auditor. Whenever I could that summer in Quebec, I worked on a novel called *The Night Auditor*. It was about this lowly accountant who worked in a cosmo-demonic hotel by the edge of a great river of time. It was some romantic entrainment about peering into the night, penetrating the night, penetrating the universe of stars, finding the meaning of existence.

I was working on this by way of getting entrance to a creative writing degree program at the University of the Americas in Mexico. Kerouac had gone there. I was not going to be distracted from this work.

In the hotel there were many hidden passageways through the service back warren, to pantries and closets and way stations, available for the staff to quickly get around behind the scenes. My friend, the manager's son, knew them all and soon I too could move about the hotel as swiftly as a *passe-muraille* — instantaneously appearing places through shortcuts and tunnels as if one could pass through walls.

There was a lot to see behind the scenes of this grand illusion if you looked. We, my young squire and I, would wander through the honeycombed chambers of the palace and people would step aside and smile politely for this young scion of the hotel manager, and *moi*. In the vaulted lobby there was an ancient arch from the old world with a capstone bearing the Cross of Malta. It was good to have friends in high places. We'd be up on the 16th floor to take in the grand view of the St. Lawrence river and the ancient stone walls of the citadel buildings. We watched the comings and goings along the great Dufferin Terrace, a promenade along the edge of the cliffs over looking old town Quebec far below. We could see the ferry cutting across the river to Levis and beyond to the ends of the earth, from our sky-master's rook on the roof in the shadow of a turret. This reached by

slipping out a dormer window in the attic and scampering up the steep pitched copper roof to narrow walkways along the roof line. Far below in the cobblestone streets you hear the clip-clop of horse drawn carriages proudly carrying the swell tourists.

As an employee of the CPR I could take my meals there which was handy, a great big breakfast of ham and fried eggs served from a steam table in the staff cafeteria. Things were pretty cool between the French and *les Anglais* there, after all it was the CPR. Though you could sense that there was a lot of unrest. The human potential movement had empowered the women's movement and the black liberation and the chicano had won their rights, as had the poor people and the gays. So the French majority was stirring for their liberation from under the heel of the English capitalists. That *was* the summer a *Separatiste* cell did throw a firebomb into the lobby of the Chateau Frontenac. I was there. It was amazing — the house detective was racing through the lobby, his long jaw cutting the air, his coattails flapping out behind. He looked very much in charge. In those days people could drive right up to the front of the hotel, just cruise through the arch under the coat of arms of Count Frontenac and pull right up to the front lobby of this colossally symbolic institution. Then just step out of a car and toss a Molotov cocktail into the lobby, where it burst on the terrazzo entrance and spewed flaming fluid around onto white lace drapes and onto a lime green couch.The bell boys and the house detective were quick to be hauling the smoldering couch out the front door into the car entrance.

In my novel I could be The Night Auditor. He was like the moon, he could feel the attraction to the sun on the other side of the world. He could feel vibrations, he was on some kind of looming love high — expanding, inflating. It was

odd, keeping track of all the minutia of tourists going through the many small departments in the hotel. I thought I could sense all the peoples who had come through this palatial hotel or walked along the terrace depositing cash and credit in shops.

We were all atwitter that summer because Princess Grace of Monaco was staying at the hotel. Our castle had been graced by the presence of a real princess. She might appear in the lobby with her retinue and make our hearts flutter with her elegant smile that made the lines around her eyes wrinkle like fine little webs of royal lace. That I would be excited about this tells you that I had found a kind of niche as a hospitality worker. It came naturally, my mother had taught us good manners. She was the same age as the Queen Elizabeth and I was the same age as Prince Charles. She collected English china commemorating various jubilees of the Queen. Mum, taught us impeccable manners compared to some I have seen.

Now in 1969 the innocent people of Canada had no clue what marijuana was for the most part. We could walk down the tree-lined streets of Quebec, puffing reefer and blow the smoke into a Mountie's face and they wouldn't have known what was up with this wacky tobaccy. *Tabernac. Challis, ca Christ.* Bob Dylan had come back from near death in the motorcycle accident, and had his great song on his new album: Lay Lady Lay. Colleen bought it. It had Knock Knock Knocking (on Heavens door), too.

The night that they landed on the moon, everybody in the world was on a high, rising on the leavening of mood that put everyone on their best behavior. That night people in bars were toasting, people were smiling and nodding at each other in recognition. Church bells rang, horns honked. And we came to realize that we had taken a first step towards becoming citizens of the universe. God the Father was no doubt standing by, beaming with pride for us. You could feel it.

Although there were some that believed it was staged on a back lot in Hollywood as a kind of bluff against the great nuclear arms race card game going on ever since von Neuman. And for some, the moon had become a prize in the race into space — another prize with which to goad production to compete against the Russians. The competition had driven the US to succeed fabulously, not quite 6 years after JFK was assassinated. Of course I suppose there was a lot of bragging going on especially back in Texas: "Eat mah moon dust, Commie." No doubt some oilionairs were thinking, How shall we carve up the real estate on the moon.

But none of that reached me. I was only a lowly night auditor in the cosmoburgeoning hotel on the river of time. Everything was fabulous. A man had flown through the sky and become a moon walker. I too was walking on the moon. I had a little job and money coming in. I had a little place and a girl to share it with. We might even be having a little child to share our lives with. And I was all right with that. And I had my little writing art, was tapping out sentences on the typer.

Yes! I too was walking on the moon but the isolation of foreign language in Quebec was getting to Colleen. She did not have the yearning for creativity to the point of obsession with art that I had. And I didn't get it that this obsession made me seem distant, distracted, adamant. I felt grateful to have a muse, and I knew even then that she is the one I would live my life with. For in the thrall of the muse, the shadow of tomorrow did not have that much of a hold on me — I was able at times to live in the now. It would just happen: your common everyday epiphany — while walking down the ancient cobblestone streets that summer. The fear of what was to be becoming of me in this life left off its hold on me. In a park or a library, I was able at times to feel a connection and agape, and to melt into the flow of now. Some other kind of ambition was pushing me. I felt lucky, and I was grateful for

the fortune that somehow, someone up there was holding me in the palm of his hands and bringing me along. The work would find me. I would get to do the work, somehow.

And now a man had leaped across the sky and landed on the moon! He had taken the first step — WE had taken the first step, out into space — the black space that we saw only at night when the earth blocked out the sun and we were in the shadow, to see the universe of stars. My god, we have sent a star-walker on his baby steps to the moon! A leaper has slipped the bonds of earth, has flown across in one giant step to the moon! It seems like the stars *were* smiling down on us. For now we had seen the earth from the moon, earthrise, planet-rise, moon-fall, against the endless black pall of the universe through which we were falling. It was a new moon-imbued world now, we were going into another Copernican revolution. Now when I walked around Old Quebec, through its portals and gates in the old wall of the citadel that fortified the old world, the smiling moon seemed much closer. As I walked over the little bridges arching over canals, or through narrow courtyards, I thought of the bridge that the Leaping Man had opened up across the sky, taking giant steps to the moon. And at night in the *claire de lune* of the noir setting of the old chateau I, moon-besotted lunatic, felt extended like some kind of spirit emanating out of the space — floating, out — a balloon-head man overlooking the river. And during the day I slept and dreamt of flying across the sky, a rainboy in curved air — flying upside down in the sunset. And I knew too that the earth is a bridge on which I walk for a while, my time here a bridge between generations, the one before and the next. Colleen's period was getting way past late and we were worried about getting married but not saying anything. I had pretty much talked myself into doing the right thing.

In the novel I was working on called *The Night Auditor*,

there was something about the teletype machine which we had in the office behind the front desk, where we did our accounting work. The metaphysical motivation driving the character was about number and meaning and "auditing" — being able to hear meaning in number like the way musicians hear sound. I wanted to hear the music of the spheres, and that was the impetus driving the story of the night auditor into that focus. I was not going to be distracted from this work. Penetrate the night, penetrate the Universe of stars, find the meaning of existence. Did I think my story was too big for me? Hell yes! But when was it ever not so. In the day I tried to write, and at night I had to put my real work away and let my paper fall and go to work and follow another paper trail.

But now there was a real man in the moon! He had arced through the night sky along a Newtonian trajectory across space and let himself land in the most alien place. And we see him on the news, taking giant steps in his one-eyed space suit, able to leap deep canyons in a single bound. And every man wanted to be him. He was the real hero with a billion faces. The Cosmic Leaper, celebrated in the Peter Max paintings — an icon of the times, had become a reality.

I was supposed to sleep during the day. That routine got old for Colleen, it was way too much stress to put the "newlywed" relationship through. If I had been more experienced in these matters of the heart, I might have been able to save it. I just didn't know. She seemed so intelligent. She turned me on to sex, and my first experience with a psychedelic. She had just graduated with a masters degree, and she had done her thesis on Beckett, and she turned me on to Borges. She had the New Directions paperback of *Labyrinths,* ordered through the St. FX book store. I was knocked out by it.

I saw in the writing of Borges a very elegant dance that he did, an ultimate convolution tango between abstract literary and philosophical erudition in the world of ideas and a spiritual quest in the labyrinth of real and symbolic objects: tigers, gauchos, impossible one-sided coins, amplified memory. I see him as a classy Argentine Fred Astair, in top hat and tail coat moving gracefully, truthfully, like a vortex through dimensions of meaning, presenting his query in a spare stylish way, always very reasonable, yet framing a place for the imagination to soar. Not to mention Borges came across as a kindly person dealing with the handicap of blindness too. The concision and certainty and gravitas in his writing struck me. He was the elder statesman of the modern. You could see his great love and belief of literature, it was for him a life practice of learning and seeing. He had devoted his life to nourishing a deep hunger for spiritual reality in metaphysics and he evidenced a great humility and gratitude, interacting with the gift of all this. The ideas in his ingenious and lucid writing were the distillations of volumes. And the ideas were wonderful. The library of Babel; Tlön, an alternative world with a made-up language; the delightful and poignant "new" attempt at a refutation of time; Funes the memorious. These stories explored the boundary between logic and dream.

Literature as Borges circulated it, was a universal, perennial, human dialog — returned to every generation. He made me want to study philosophers, so I could follow the continued fractions down into the infinitessimals of Zeno's Paradox, or appreciate the mystical rational synthesis of Spinoza, or catch more of the allusion and level of story, the symbolic matrix, the game that Borges had going on — pushing a theme along through his content. He was encouraging in the world of new science to take on these ideas and know what people had come to know. And as

all good hippies do, I wanted to feel the machinations of a hyperdimensional object, undulating in and out of our three dimensions. I definitely needed to study and learn more!

I began to spend what little off hours I had in the reference room of the beautiful English Library of Quebec City, browsing the Montcrief translation of Proust, or the Catholic writer, Leon Bloy or writing poems in the style of Gerard Manly Hopkins, about spirits moving through the flesh. This most beautiful ideal of library, had a grand salon reading room. It was cavernous, with walls a robin's egg blue that opened up past two stories to the domed ceiling. A walkway went all around the perimeter at the top of the first floor leading to rows of stacks. Tall windows opened, to admit a summer breeze off the St. Lawrence, to loft the stylish bright yellow curtains, which were regal and Provencal with blue ornamental details around the border. The building was classical, elegant. This ancient edifice transplanted from France was inside the old walled citadel of Quebec, behind gates that separated it from the modern city.

But during the day I had to get some sleep to get ready to be up all night for my job; I never got used to it and at times could be dragging my ass from fatigue. I told myself it is not good for budding young love relationship but I was the only one working right now, and I needed to bring in the rent and groceries. I was but a foreigner and damn lucky to have a job, especially in that castle which had captured me in its spell. And so I went to bed in the day or rather, we went to bed — the three of us, my girlfriend, Borges and I. For one day she was there beside me reading in the middle of the day, a Borges story and when I was awakened out of a light sleep, I told her what I was dreaming: "It was about being underneath some stairs somewhere. It was down, down in the basement of some person's house that I didn't know. It was like my grandfather's cellar, down wooden stairs, with

piles of wood and coal and shelves of pickled vegetables and jars of home made jams and jellies. It was Some Person I didn't know."

And she said, "Wow, that sounds like what I was reading here in Borges. Right her beside you."

Then she read a passage about some guy going into a basement and encountering a collapsed dimension in the shape of a circle within an triangle, an all seeing eye, into which he could look and see many things going on all around him. Some might call it 'psi-ops.' (At the time, both the Russian and American espionage communities were doing experiments with application of hallucinogenic drugs and truth serums and other drugs, as well as paranormal remote-viewing and extra-sensory perception.)

We were astounded at this coincidence or synchronistical convolution of me dreaming what she was at the same time reading and had a special bond after that. My relationship with Colleen was so significant that it felt like some kind of telepathy was at play in the fields of destiny. Or maybe it was just the looming into this love high I was undergoing. But from that day forth, he haunted me. The three of us were in a mystical cabal, my girlfriend, Borges and I.

The next day, I decided I'd been hypnagogic, half in and half out of sleep — the night job, was driving me crazy. But in fact it was the beginning of a fever. The fiction of Borges, the *ficciones*, was metafiction — fiction about *being* fiction. It was communicating a story and also was a story about how we know it is a story. His literature had gone self-epistemological! This was a kind of "philosophical science fiction." Borges explored ideas like they were features in the landscape of Mind. He explored ideas in the context of the original creators, so that you could see the poetry of discovery, in context.

In the beautiful English library of Quebec I searched

for other Borges writings and found his book of essays, and other writers he mentioned. I was delighted to find he had written about J.W. Dune, who explained about a parallel universe in a different time dimension into which one might enter in sleep and dream. I thought one might also enter it because of the strength of his emotional compact with his girlfriend — that the emotional right brain, containing a primitive intuitive consciousness suppressed by centuries of evolution, had been awakened by love and by literature in a process that takes place during sleep.

I was reading Proust and Pascal, struggling in French. I don't know how to say it, I was awakened. I felt real and alive; I was a part of what was going on. I was in some kind of flow. It was vary poignant and a little frightening, because this precious bracing attention could be easily taken away. I might find myself looking at the movement of the wind through the grasses and reeds was like looking at some great magic painter man working, like an invisible van Gough. Things were brighter, sharper. My senses were more acute, awakened. Did I have an overactive occipital lobe? Was it serotonin cascades in the frontal lobe? Or was it just love?

I kept writing, the story of *The Night Auditor*. And it just wasn't nearly as good as what I saw in my mind. The Night Auditor was about this lowly accountant who worked in an omnipullulating hotel by the edge of a great flowing river. The rooms in this hotel were spaces that could expand and contract, could bud like crystals. It was about extension and closure. I didn't have the words for it then.

In Quebec, with my night auditor job, I had become a contrary man, sleeping during the day, and at night being up working in the chateau. During the day I had strange dreams; and at night I looked at tables and columns of numbers, following the receipts of ghostly guests in the chateau. During my break I would wander around in the great edifice

looking out the windows at the moon on the river. And the mysterious fever persisted, through days clouded by the fog of sleep deprivation, leading to hypnagogic dreaming and attempts at automatic writing and sketching, and somnambulistic walks through the intricate stone and brick labyrinths of the old walled city. The writing was about the projection of a vast desire into the night, the day dreams were about writing and the nights were spent wandering the hotel with everyone's dreams rising out over the moonlit river.

    Still no sign of Colleen's period; we were trying in our separate ways — secretly, to get used to the idea of getting married. I was in theory, OK with the idea.
    She smiled with consternation in her eyes. She bought some new albums. We had another young guy in the house who kept normal hours and he seemed to be spending a lot of time with her. They liked scoring and smoking dope. He was funny and sweet, compared to my tall, dark, grim.
    I was committed to getting this book written, and coming to understand something of the Night Auditor. I was committed to working on this novel that somehow allowed me to see who I was, who I was becoming. And too, it was supposed to get me into school — furthering my writing career. I was blinded by the light of possibilities, not monetary reward but of the act of being able to engage the forces of the world that brought us here and conspired to mold us. I was dizzied with the possibilities of the interaction between dream and narrative and imagination and how all this could be amplified with literature and drugs. The diagesis of Borges, in the logic of the moment of the text, moves not from raw experience of the actual, for we are experiencing the signs, the map of some one else's experience, which itself may be the experience of someone else. The semiotic moves from perception to memory where

persistence is dependent on the power of experience. This is used again and again in comparing patterns across prior experience.

It is into this existentialist division between sign and *Dasein* that Borges draws us, through stories and parables and constructions and sophistries and savories and poetry. The labyrinth of libraries and books, gauchos and knives, mortality and the infinite, the absolute oneness and the constant separation from it, seduces us into finding that Other voice within ourselves, the one that is awakened at the moment of epiphany. These objects in Borges are symbols, signs of something deeper, of things seen in rare epiphany — that moment when coincidence has its fortune to give out to the mind, to keep it moving into appreciating its workings in generosity.

Under the influence of Borges, one day I looked into a mirror to see if I could see this other me operating behind my eyes, the windows to my soul. You can't really see the reflection of your own head looking into your own eyes in the mirror, unless perhaps at some angle where the projection was just right. But just the act of focusing on the reflection of me looking back from the mirror trying to see my reflection in the eye gave me an uncanny feeling of parallel mirror worlds infinitely going inward toward some kind of archetypal self. It was excruciatingly embarrassing though no one else was in the room.

We had those parallel mirrors in the lobby of the Chateau at the elevator landing. And this sense of the other looking through my eyes began to follow me to work. He doesn't let me see him, for I can only look through my eyes. Oh I have tried to capture him in a mirror, but he always goes into an act, just like me. You would think he would be visible in a candid photograph, and perhaps he is, though there are no photographs of those days, certainly none that are not posed.

Now sometimes in the confusion of sleep deprivation, I would forget which one is me. It is as though there is one of me for the day, and another of me for the night. Or one of me for the workaday world, trying to sustain my existence, and the Other of Me, looming out into its birthright — to understand its purpose among the stars. There was a residue of *deja vue* left over in the contrary man that passed through my strung-out personage. I told myself it must be the effect of the sleep deprivation from wandering the night and trying to sleep during the day, trying to reset the circadian rhythm of my body clock to being a *doppelganger* man on the night shift. I might find myself singing the Christmas jingle: 'He knows when your are sleeping; he knows when you're awake.' He sees through my eyes, I am an organ, an extension, a waldo with which he handles matters in the outside world. He's got my heart, me my emptiness. He's got my envy; I've got his longevity.

I had become a person of the night, was out and about when the rest of the world was asleep, dreaming. I'd watch: — the Night Auditor peering from his towering perch on the 16th floor — in the river of moonlight, as their astral bodies or souls came floating up, out of the windows of the buildings, floating across the moon lit, cloud scudded night sky, each projection on a thin silvery thread. Yes. At night in dreams, the soul is held to the body by a silver thread. It is attached through the fixed point, coming out of the crown that swirled around on the back of the head. The soul floats up out of the self, as the dream narrative unwinds into the pearls of coincidence presented as perils constructed by real or imagined influences. The soul does not come out every night, though it is not bothered by the weather. Only the whether . . .

Some say that there is a field-effect of communications between dreamers. Colleen and I were oeironauts in our little

bed boat and we did share our dreams in conversation. The soul weaves the Subliminal Opera of Dream Lands into Time. In dreams, time is tachistoscopic and tardiochronic — the experience of great long times compressed in a moment's flash, or a moment stretched out into slow-mo sub-frames of continued fractions of infinitesimal detail.

The soul knows when to come home to the body, though it is most active trying to finish up the night revelation in reverie at the last moments. So, often some of the exposition gets carried over into the hypnagogic state — in the between time. Then consciousness comes back and the soul goes under in the sweet slumber of waking.

For him (the Night Auditor) I am a cipher, an operand in a computation, an argument in a function or a disputation. I am an automaton acting out an unfolding narrative that the universe is creating in the subliminal opera of my mind. I can only see out of my eyes, I cannot see who is looking through my eyes. I understand I am not supposed to see this Other because it could create dimensional dissonance if I see who is looking through me. We are born out of uncertainty and action, from these we are constructed, it is how we proceed. We are born to be observers to the universe's unfolding, it adapted itself for use, it adapted itself *for* us. And as I see myself it sees itself through me. He knows when I am holding back, stuffing stuff into my infinitely expandable occluded side. He sees it, he knows it, he is at home in the realms of uncertainty and shadow, chaos and chance and I am disappearing into the obscurity that surrounds my life. I will be known in that moment when I disappear into his eyes, the moment my face disappears. I will be following behind my own eyes when I am able to shut off the ego and be born into the now moment, to float up over the hill and off into the blue.

Well, look into my eyes, they are as blue as the skies. I have my dreams too as you do, and we are all here part of this

great Other that supports us, part of the unfolding. There is so much I don't know, I don't understand, in my little time line. We are all part of this together in our lives. Sometimes the other has my day and me, I have my night. But sometimes, I have my days, and my nights are lost because I do not dream.

I wonder why the one who runs around during the day, is given so much more credence than the one of the night. It's as though it has to see the day through my eyes, so that it can show me to myself in the night. For I take my dreams away from the night and go to the day on the other side of night. That way we Switch — I see in the day, and he sees the night. It is somehow involved with literature — the narrative, the construct of who we are, for the world is constructed for that who.

Colleen and I were close, intellectually. We had a great interest in writing in common. So close, I would sometimes be dreaming what she was reading there in the bed beside me. Just to say the second part of that, 'in the bed beside me' — it was so intimate. The reading passed through a mind and reflected off a mirror there into a dreamer on the bed beside her. The little room in the old city overlooking the quaint old-world street below, was like a point in the stream of the world outside the window through which was surging consciousness. Have you ever had the experience where you are so in the book that there was no book and you were in the interior world of truth? It was intimate to have this person beside me quietly reading. Taking in the meaning through the mind being spoken to intuitively through the imagination. The artist there confronting the status quo, putting the people he loves in the picture as they are — young — in a world into which we wanted to dissolve like the sunlight playing on their faces through the leaves. Democratic, not elitist. It is

not so much about what they presented as how they captured what they saw. That is what fiction is, the wide swath fanning out into a world of possibilities that the truth cuts through the mind.

The situation of not having a job began to really bother Colleen. She didn't want to be dependant on me, though I was OK with it. She had a lot of time on her hands and the isolation of not being into French started to wear on her.

Finally, my girlfriend's period arrived — 2 months late.

We looked each other in the eye, and at that moment I think said good-bye to what might have been. A parallel forking path, had been closed to us, and we were I think, grateful. I knew I was.

She went down to Montreal to visit some friends in some crash pad. I was supposed to eventually follow her. She decided to leave me and go stay with her friends in Montreal.

I went with her to the Gare du Palais CPR station to put her on the train and say good-bye.

I was sad to see her go.

Then I was on my own. And I started slipping.

It starts when you wake up. You reach for her but then realize again she's gone. She's really gone. You head out to get some breakfast, the streets are bustling with foreign chatter.

Back again, you sag into the armchair, thinking: She can't turn me into a mess.

How did things get like this? Just a few days ago things were OK, were cool. It was fun getting high with Colleen and the other borders, making fun of our landlady who we called *mille fesse* because her butt was a mile wide, waddling down the hall. Now, Colleen's voice is replayed in your mind as you try to find where was she so unhappy. You try to explain her abrupt departure. Why? as the event plays itself again.

"I've got to get going, got to get a job or something." Colleen had said.

You were pleased, but she said, "I'm not going to be able to get a job here, I've got to visit my friend Marcie in Montreal. Then I might go visit my sister in BC."

"I'm sorry to say this, Walker. I have to move on — out of Quebec, you can stay here without me, if you want. I've tried."

You recall that moment looking each other in the eye, realizing the pregnancy had become a moot issue.

You end up thinking she wouldn't have left you had you promised to marry her. You were let of the hook around marriage and relieved. Still now she was gone and you are alone.

It would be cool to visit Colleen in Montreal. She invited you, but you were too stubborn.

You start another letter to her at the writing desk across the room, in the light of the gooseneck lamp. You don't know what to say. You ball up a page and start to write again.

You never knew you could miss another person so much. It gnaws the heart like a mongrel scavenger.

It was nice having a woman to lay across your big brass bed. Her warm body and lovely breasts all yours to have and hold. It was wonderful when she moaned under you sometimes. And her intelligent conversation and philosophical hunger. No pretense here. Or so you thought.

You aren't tight fisted by nature: the problem lay with her not having a job and you having to work all the time. You get so bored and confused if you can't get a little writing in once in a while. And it is so hard being an outsider in the language, you just can't jive and joke in French like they can. They treat you like some kind of retard.

You sigh and flop back into the chair. I could join her down in Montreal in the commune; even go out to

Vancouver with her, never been there, it would be great to go with someone who knew someone. But I have to finish this book so I can get into a creative writing school, get moving on, in the world that I want to move in.

The hot winds of Quebec summer move the yellow curtains and you are mute at the table.

And then you think: I'm 21 years old, not some weepy kid! I am not going to take the long bus ride to the Isle d'Orleans bridge and jump off into the swift wide river over some girl.

You get the bright idea to quit your job before Labor Day. Must give notice. Want to be able to work for the CPR again. Bye-bye, magical castle in the sky! I'm no suited-dude, bean-counter for the conglomerate! You are slipping into an attitude; not sure what kind of job you'd take on. Anyway, you feel confident that taking this action might alleviate the sadness which is taking over your life.

Now when your walk around outside, everything is amplified, the world is wailing at you. Something had changed with my brain. It opened up. Filters got removed.

Thereafter this Other Walker began to think of himself as the Night Auditor, the mind behind the arras, the mind of Night itself, the fabric of the dream world that everybody was in. I was the Night Auditor peering through the veil at the machinations of the universe. Maybe it was some kind of after flash from the psychedelic experience. Maybe it was the effect of sleep deprivation from having a night job, or the social stress of being intimate with a girlfriend, or maybe trying to write my first book and not having any idea how to go about it, but being determined to do it. All this was stretching my mind quite thin. The fabric of the surface of my reality was being stretched thin by the cognitive dissonance within — poking through the cerebral skin, I was becoming transparent. I was having strange experiences,

with the world sneaking up on me and giving me these partial epiphany experiences about the underlying workings of the universe beneath the surface of the natural world.

The Night Auditor could pick up signals from across the galaxy. He was a dog star man. Seeing the earth from space he realized how truly alone and vulnerable we were, in the vastness of space. We were orphans left here; everybody sensed of that. And we are all brothers and we need to take care of each other.

She was my guide on a first mescaline trip. She took my psychedelic virginity as well as my sexual virginity. I was not ready, though I postured — acting more experienced than I was. But the mental constructs of my reality had been seriously dissolved, and I was struggling to find a place to stand. I was not of this earth. Before I felt things so profoundly. Now when I walked, I could have just as well been on a wheel moving the earth under my feet in an endless squirrel cage.

Before, I was high on love and the idealism of the setting. My heart was lit with love — fully consummated love — for the first time. I felt things so *profoundly*, — the night and the moon. It was the moon. That's it. I had been a loon in the *lune*. *It* was no longer a foreign, idealized body, it was another world, somewhat like our own. The moon was born when the earth and another planet collided and the crust of the earth got ripped off and flung into orbit, while the core of the other planet sunk into the earth's core and merged. Now we have a planet with a outsized core and a thin crust to keep things nice and warm. The incoming planetoid whose core had merged with that of earth in the collision, had ricocheted — along with the earth crust it had blasted off — out into a circulating debris field. The planetoid became the moon as it lifted up the flung-out old crust of the Earth and swirled it around itself in a sphere. Thus the earth with

crust removed now had basins for the oceans to fill and a much thinner crust that could float around in plates, pushing up the land. So that it wouldn't be just a total water world. And man could walk upright and navigate space and use tools and wonder about the universe of stars in the night sky. Now when I looked at the moon, I saw it as a luminous being, wearing the skin of earth.

It was Love, that you had been feeling — for the first time. And you had fallen out of grace with Love. You had been learning to float in the weightless levity of Love...

Some kind of boundary had been crossed or had broken down. The thing that filters out the world, that organizes the world, had been made more transparent, had been breached in a tidal wave of sorrow and resolve. You were going down the drain, being dissolved in a solution of sorrow, and your ego had shut down.

The world was broken. You are trying to hold your form together and it was coming apart, for all things are the result of attraction and repulsion. You had fallen away from Love's attraction and the further you got, the more repulsive you became. Breaking up with girlfriend drove you over the edge and you fell deep into the well of your self. Or maybe it was that you were having these flashback from the mescaline trip you had done on the Plains of Abraham. There you had been drawn into the green world like never before. And now certain cool hollows in the green world were calling you back. You had become a contrary man; nights had become day, days had become your nights. That must be it: you were sleep deprived, I was walking around in a dejected desultory hypnagogic state and my dreams were spilling over into my waking life.

It was horrible and it was beautiful; you were in charge and you were a plaything of chaos, terrified. The world was moving within and without you. You could transcend your self and float out and be in a kind of unity with the motion

— the forces there that were moving on the wind, shaking the trees, making the flags and clothes on the lines dance and oscillate and try to fly off and escape. Or even the forces that moved the dark clouds and the rain in.

I was wandering around asking question of things as if they were sentient, as if they could some how answer. I asked the bilious clouds: What makes you different from the high wavelike the altocumulus, from the great thunder heads. Or I asked the trees: Was it the swaying in the breeze that built you up like muscle under a shirt that causes your bark, to burst, to split in folds and furrows running up and down your length like that?

It was something wild in a young mind, to be privy to the workings of the world like some great swirling light show, or flowing expansion, or a dispersion of particles each throbbing on strings connected to the sun.

It was a crazy time, hot summer days of bright light, but then the air was so filled with potential rain, and it was just looming. Something was just looming, about to go off. And everything was so green and large this summer in Quebec proliferating, climbing, grasping trying to fill out space with its progeny, trying to situate itself in the best possible light to win the struggle — everything else be damned. I could see the suburbs from the Chateau filling the map with their streets and blocks and driveways and alleyways. And parking lots and malls. It was the expansive accretion of the city, the layers of habitat build on previous layers and families. I could see the land of the great river valley dropping off to the water and I could feel the river in my blood, for the veins were like trees and trees took up the water held in the hydrological cycle locally by the river and it was all so miraculous and phase-locked in. It made you want to get down on your knees in weeping gratitude and acknowledgment. Or sweat out the river blood in knowing that these

rivers that flow in animals and plants and the geology and me, shape and are shaped in co-evolution >> it was an almost Platonic idea of noumenonal pre-existence like the circle or the self-reproducing automaton, or the matrix or the game, or the colonization of a substrate by a colony, or the diffusion of perfume across a room.

I had been reading a lot of Pascal and picking up some of the metaphysical terror looking out at night into the lacy star fields of the stochastic universe and feeling that fear at the edge of things where your radical doubt is amplified by aloneness into a laser-pure anxiety undiluted in the possibilities of distraction by the Other as the fear becomes a constant companion haunting and shadowing every thought.

Or maybe it WAS just a flashback from your recent mescaline trip. So you have slipped into a dissociative, perhaps psychotic experience, what you started to call the Green Fuse. You had this sense that the plants and trees around you were alive and aware, and in touch with the life force and were "looking at you." It was frightening. It had something to do with this big patch of horrible looking weeds in a vacant lot. The weeds with their jagged edges looked like the mandibles of insects, or worse, some kind of cancerous tentacles. Your brain tried to process it as just a weed. You tried to assuage your anxiety by telling yourself: There is not some great green thing in front of me, some Jolly Green Giant of weeds slouching across the landscape of your days. And laughed it off. However the image persisted for what seemed like a very long time, weeks. You didn't have anybody to talk to about it. Going to a psychiatrist would have been the last thing you would have done. You told yourself it was some kind of an externalized obsession. You went about your cognitive business, brushing aside what was perhaps a outward sign? A nouveau catholic mystic might say the weed was some kind of rebirth symbol,

something that they kill and eradicate but still keeps being reborn. Or perhaps it was just a symbol of you, a weed in this foreign landscape. This edgy destructive remorse-imbued paranoia was a very powerful and frightening experience. Unforgettable. You didn't understand it. But in the end there was no fighting it, you knew it was something you had to surrender to, a power. You tried not to think of it.

It ran a whole lunar cycle.

I have never felt like that before, I couldn't bear living in the apartment, with the hip young border asking questions and all, so moved to another rooming house. It was in a house with a bunch of young French guys, who probably thought I was too serious. They introduced me to some cool French Jazz and Bach. I quit my job and lost that apartment and was drifting into chaos, in the foreign summer's air with the plethora of incomprehensible language rising like a thicket of hindrances all around me. For a while I would still go up and eat breakfast at the CPR staff cafeteria, until they asked me to sign some kind of registrar. More and more the wolf at my door was getting ready to bite me in the rear. I had to shift around from rooming house to rooming house in the dead of night cause I couldn't pay rent. During the day I walked for blocks and blocks as though something was chasing me. I just wanted to get out of my head. Into what?

I kept going over the telepathic dream at the intersection of Borges and I and Colleen. I am going down into some place underground. My head is floating out the window. People are walking outside a basement window. There are glass jars of preserved things. In the glass, is crystal, in the crystal is matrix, in the matrix refraction off planes, things come and go on the planes like movie frames. It is something being projected from all around. It is the world itself.

I was being inducted into some kind of Secret, that I was not supposed to see. Why me?! Who was I to be given this.

It was pulling me apart. My face and forehead was being jacked up and a whole new entity was being dropped behind the façade. I was being taken over. I had to resist.

I had to sweat out the coming fever of alien possession in the river blood of the *roman fleuve*, coming in, into this foreign place. I should have gotten into some kind of a group therapy, or gone to see some kind of psychotherapist, but felt nothing but stand-off vibes coming at me a million cycles per second. High frequency passing right through you. But then I would take heart: we had been given the moon!

I am some kind of ecology which sees itself in the dream and in the day. I can travel out of myself and see myself from outside in the dream. How is that possible? What a gift!

Gradually my circadian clock got synched up.

I was trying to decide whether to go back into college or go out into the world and discover it more on its own terms, a world that was coming in to me, these days at a terrifying rate, that I could not stop or control.

You must remember to maintain and walk the line.

One day in a fit of despair I collected up all the many typed pages of my fat novel and filled the bathtub with water, and slowly dealt them into a watery grave. I watched as the ink blurred on the page. Soon my eyes were full of tears too. Sometimes I wish I hadn't done that.

I saw my girlfriend Colleen at the commune in Montreal. We were friendly, respectful; we had been through a lot together. She tried to talk me into going to Vancouver with her, but I had to get back to Texas and some people that knew me. I drove her down the Trans Canada highway past Toronto until I had to head south to Detroit. And she continued on to BC. The last I saw of her, she was standing on the highway heading west.

I eventually did go back to school, but changed my

major from English to Engineering and Math and eventually got a degree in Physics.

The details of the months and years that I, or we, Borges and I have lived since that crucial, fateful summer of 1969 when I fell under his spell into a psychotic break in which floodgates of the metaphysical world were opened to me, have never really left me. It was a fortunate fall, one that made all the difference in the world.

I came back to Texas, having been accepted at the University of the Americas in Pueblo, Mexico to study creative writing. I had obtained the necessary letters from the police, and from the parish priest, but couldn't raise the money. So I decided to study science instead at the University of Texas in Austin. I got the money to go to school from student loans and graduated in 3 years with a B.A. in physics, after having started out as an EE major. I found engineering too careerist, too much nerd macho behavior. Though as it turns out, circuit theory was a great introduction to topology and scattering; as was Boolean algebra to vector spaces. And the metaphors of computers and programming were causing a paradigm shift in the imagination. The world was a brain doing a calculation.

I had no girlfriend during that whole time, and really busted my ass in college. Differential equations, group theory, a very heady intellectual rush. Toward the end of it I met Diana and we started living together. We made a trip to Mexico in 74. In Austin I started teaching electronics in a trade school. I eventually got relieved of my duties for growing a beard. Diana and I broke up because she wanted to marry and I didn't. Somewhere in there the Vietnam war ended.

I began the long slow sinking down into the body out of the head, becoming a physicist in recovery, something that took years and years. Along the way I slipped further down the downward path to becoming a writer.

Letter to a Campfire Girl

Laura was going away on a camping trip with her high school senior class. It was supposed to be a bonding group-experience before they all scattered to the wind after graduation.

I thought I might write a little letter to her, tell her how fine I think she is. Perhaps maybe even come off with a poem. It would have to be cool cause no doubt she would show it to her girlfriends. Don't want it to be something they could get all embarrassed about and start teasing her. But maybe something to blow her mind while she was out in nature getting all trancendentalized and passionate about her fellow students. I thought to sneak it into her kit somehow so it would be a surprise, somewhere down inside her stuff, maybe tucked someplace down inside her pillow that she might find it when she was rolling out for the night. I got started:

> Dear Laura,
> This is just a note to say, I miss you.
> Got to have more of your sweet self around here.

Then I thought it might go into some kind of a poem:

> Girl lately, just thinking of sweet you, puts me over the moon. / You're the first thought I have in the morning / and the last one I have before sleep.

God no. That's no good. Gotta hot it up with something.

> Ours will be a love that lasts all time / Shimmering and dirty and wonderful / hot times in a dream circus. . .

Circus? No-oo! That won't do. It's funny though.

> How would you like a daddy / to live in a circus tent / to help you out of your body / and do it all day long on a trapeze net /And never have to pay no rent.

No, no! Too vulgar. Gotta get something more classical, PG — that both she and her girlfriends will like. Some Attic grace from Sappho maybe. . .

> Venus in blue jeans, child of Zeus, / with your neo-hippie / pre-Raphaelite wiles, / you're an answer to my prayers, / if by prayer we mean the articulated desire / to feel myself through you.
> You come around knocking at my door / bringing all kinds of pleasures for me to adore / then I walk you home around midnight / to your father's house . . .

Then maybe I could ask: How would you like a daddy? No. NO! Can't go there.

Here I am 30 years old and consider myself a writer and can't even come up with a decent love poem to the modern young woman of today. I just want it to be something straight from the heart. And I need to tell her that I am too old for her.

I need to outline: Tell her how much her being in my life has made me a better man. Say something about her struggles with identity at school. About the sad story of her parents divorcing, about the responsibility she feels for her sister. . . about how proud I am to know her, that she is doing a great job with her writing.

Image: It will be so nice to be up on the swing in the tree house perhaps drinking a little beer and thinking about just staying up all night and teasing and having fun, and looking and howling at the tranquil moon peering down on us from infinite time. We are blessed that we have this time with each

other. It lets us be touched with the infinity of love . . .

Maybe I ought to bring up the age difference. I wonder if it means we have no chance. Probably does. She is going to be wanting to get together with these young guys.

> I just wanted to say something about age differences. How are you getting along with that?
> The psychologists Eric Erikson talks about the 8 stages in a person's life.
> I cribbed together a conjecture about how these ages are always still around in a person — a personality. And about how the soul looks through the personality to play with the spirit. The previous stages go underground. And are part of the being, out there like transparent surfaces protecting the being, like the petals of a flower. Or the husk of a nut. Like the skin of an onion or an artichoke. So this is sort of the artichoke theory of man.

Artichoke! Gag me. Soul being the various stages of life's passages. Make this about the hunger of the soul for love.

> I say that the soul grows with the life of a person.
> It goes through passages where it takes on more
> understanding, while keeping what it had before.
>
> The soul is looking for love.
> In the newborn stage, the soul starts
> learning trust and mistrust.
>
> The whole universe is compressed down into a single
> binary being, the mother and the infant.
> Who could remember that time, it has gone unconscious. . .
> The newborn thinks: Yesterday you came over to me
> and the day before, and the day before that.
> Trust came in.
> I am here.
> Don't cry,
> I open my blouse
> Hush, shh . . .

In the toddler stage the soul starts to feel its way around
space through the personality of the host.
On the floor, one struggles with autonomy.
Standing up I spoke
in the world of giants: I too exist!
And: No! I won't!
Say No. Confound! It is the beginning of ego
that will sustain you.

In the kid stage it is more autonomy . . .
In the school as before,
They spoke:
The boys go in this line
and the girls go in that line.
But this desire to gain power is also
haunted by ghosts of guilt and shame
 – "cooties" they call them.
We will whisper secrets to our same sex friends
and listen to them.

The 4th stage has you in school, a middle aged kid,
It is about industry vs. inferiority . . .
So exquisite the quest of efficacy:
It can't be too difficult as to overwhelm,
nor too easy as to not provide growth.
You make a covenant with your Muse here.
I wish they taught mysticism alongside algebra and music.
Where the soul comes from and where it is going,
cannot be expressed in words.
In this stage as in others the soul asks:
If I do this right will you love me?
Is this how love works?
The soul is looking for love,
it struggles with trust, with belief, with the ego, with growth.

Sorry I got going a bit on this riff, but hang-in with me.

The current stage is Adolescence. It is about identity.
The four remaining stages take what the child was given

out into the world,
the youth — adolescent stage, is about relationship to peers, and the sense of continuity in your identity — no gaps here.

This is where your first transference experiences in love occurs. The ancients depict Love as a fat little cherub with a bow and arrow. It is your infant self, looking out to see who it can make fall in love with it.

The next stage is Intimacy vs Isolation, young adults in the word of work and dating. You are going to have so much fun.

And the later stages . . .

The soul drives the personality out into the world for love.

The soul speaks through the personality:
"Love, I'm scared that it will not be
like what I expect, what I am used to,
what I have learned in family and TV."

The Soul will whisper secrets in your ear
Calm down,
Pay attention
Let your being float out and be with the other.

Wow I'm really goin' out there with this aren't I? I can't be telling her this. But just for myself I want to know. What do these stages of personality tell us about how the soul seeks love in the world, the soul is of love.

The soul is looking for love in all of these stages,
Looking for love in all these places.
Love looks different to the soul in all of these phases.
The soul brings all of these stages to every encounter
to feel with and be felt by other souls.

I love you because you help me feel my soul,
as it feels vulnerable and in your care,
as it feels defiant and enslaved and nurturing too.
You can feel alone and together
and capable and treasuring like a parent.
I feel myself more whole, in love.

When you appear at my door with your smiling face, I think of how touched with grace I am. I want to change myself into a mirror so that you too may feel this grace.

Ye gods and little fishes! I was trying not to be too hip and cynical, and now have devolved into some gibberish sprouting spiritual seeker. I've got to watch it, or I will become a wonder-infatuated bliss-ninny.

I couldn't put all that into a letter, it would freak her out. So I just tossed off the more or less conventional sentiment. I couldn't manage hiding the letter either, so just handed it to her in an envelope that said: Open when squared away.

She waited until things were quieted down at camp. It was just her and Wendy hanging out in the tent when she opened the letter to read it. She told me later, "It was raining and we were miserable. Luckily Wendy brought a little smoke. An I had brought a bottle of wine too."

Dear Laura,

This is just to say how much your being in my life has made me a better man.

I am touched by your life — your struggles with identity at school, the sad story of your parents divorcing, and the responsibility you feel for your sister.

How proud I am to know you!

I thing you are doing a great job with your writing.

It will be so nice to be up on the swing in the tree house with you, when you get back, perhaps drinking a little beer, and thinking about just staying up all night and teasing and having fun, and looking and howling at the tranquil moon peering down on us from infinite time.

That we have this time with each other in love — lets us be touched with the infinity of love.

kind regards,
Walker

Tales of the Wild Seed Women

Under the influence of love, and happiness in relationship with Wendy, Wild Bill started writing a new novel to be called *Tales of the Wild Seed Women*. On our walks or at the kitchen table, Wild Bill would be on the trail of a tale, working it out.

It was an apocalyptic story set in another time, in a world that had been bombed back into the stone age, a future after nuclear holocaust where people had to return to survival by sheer physical prowess. I don't recall if the main character had a name, he was just the Noble Savage. He is the scion of a chieftain and he shepherds this small nomadic hunter/gatherer tribe in a landscape populated by savage predatory bandit peoples of strange cults and practices. In this world there is a profusion of animals that resourceful nature has replenished. It is a landscape where the cities and civilizations are ruined, but where there are useful artifacts in the rubble. It was a very realistic portrayal of those times that lurked in the back of all our post World War II boomer minds — for we were raised on Cold War, in the mental atmosphere of pandemic nuclear dread — preparing for a future that could be gone in seconds. Indeed the naive grade-school duck-and-cover drills where you crawled under your desk and cowered, scrunched-up on your knees, nose to tail with your fellow student seemed to suggest you would have just enough time to kiss your neighbor's ass good-bye. We grew up with the idea that there might not be any future at all.

In the future world of this novel we are in a kind of pre-historical romance / adventure fantasy. But it was realistic — Wild Bill was gaming out survival scenarios, like they did at the Pentagon. He'd write it with a Bic pen on yellow legal

pad and then type it up. He wrote an almost letter-perfect first draft. I think that is inherent in the idea of Tale: it is an oral text composed in the mind beforehand for telling, and then dictated out whole to the page. Moreover, like the storytellers of old, it was visualized, and described as though present in real time. I would encourage him to read me his sheets, and he obliged. Or sometimes he would just tell the tale from memory as part of our discussions on the dog walk. Our many discussions, as we walked along the river in the shade of tall trees on the warm Texas days, though usually about women, were also about the half-life of radioactive particulate fall-out, or improvised weaponry — making shanks, bombs and traps. Or about gangs and warlords — but mostly it was about women — and the spirit.

The *Tales of the Wild Seed Women* was a very thought provoking project, and it occupied much of our time together in the coming months. Wild Bill and I talked so much about survival scenarios that at times I felt like we were living in one. Being part of the story helped me not only understand nuclear dread, but also an alternative, more hopeful mindset. I was helpful in the tale's unfolding. What I am telling you now is recalled from those storytelling occasions. The manuscript that Wild Bill was writing is lost in all the unstable shifting about in the living situation.

The Noble Savage grew up in the aftermath of the world wide apocalypse. He was with his father and mother in the Sangre de Christos mountains of New Mexico when the bombs hit. There was a short nuclear war. Some glitch in the computers, made one of the nuclear power countries think it was under attack from another. Then the use-it or loose-it admonition of mutually assured destruction automatically kicked in. His father was a geologist and his mother an archeologist. They hunkered down in an abandoned silver

mine with a few friends and colleagues who happened to be visiting at the time. They got a few minutes warning before the nuclear missiles started knocking out cities. They just did get underground before great clouds of radioactive fallout circled the globe. They quickly got used to using the Geiger counter. Batteries were a premium. One has to avoid getting the radio active particulant on your skin, in your breathing or in your food. They laid up in the mine for 10 days, until driven out by starvation. What they found was a world turned into chaos. They looked for unopened cans. And so it began: Survival in a lawless time.

The electrical grid was out. The phones were out, no radio broadcasting. A ham radio operator might still be up if he had a battery back up system that could be crank charged. There were few solar or wind power generators.

The group was able to find and have side arms.

We talked a lot about weapons.

I said, "Yeah, once the electricity goes out. People gonna know that they are not in the 20th century any more.

"Food production is totally based on electricity, to pump water, to process. And the computers will be down, and there won't be fuel production so there won't be distribution.

He said, "Gangs will take over areas of the city. If you are not gang affiliated, or part of a group, you are going to have to become a refugee. They'll be on the move in streams, clog the highways until that first tank of gas runs out. Then they will be on foot. Bands of marauders will stalk the countryside — raping, looting, murdering, clashing with each other. Of course, it is a wise gang leader that doesn't get his crew killed; they will avoid clashes if possible unless they think they can win."

I said, " Wouldn't the government restore order? There's the Army, the Navy, the Air Force, . . . the Marines. The

Coast Guard. Presumably they represent the establishment and will try to restore order. Certainly they will move to protect their stockpiles."

"But an army runs on its stomach," he said, "and with the electrical grid down, they are sooner or later going to have to send out patrols to start foraging. There are a lot of armed groups in America. And they will certainly clash.

"I read where there is a gun for every man woman and child. 200 million."

"Yeah. What about the Border Patrol. Maybe they will side with the Mexican cartels, the Sinololas and the Zetas or the Gulf Cartel. They have lots of firepower, they have their own military training grounds."

"Yeah, could be. Maybe race will win out after all."

"You know it will."

When we talked we sometimes got into one-uping each other in terms of the gallows humor of the survivalist.

"There are a lot of Mexican gangs in this country. The Latin Kings, the Latin Disciples, the Mexican Mafia."

"I even heard of one called Hispanics Causing Panic."

"We had one in San Francisco called HHG, which stood for Happy Homes Grandé. I don't know where they were going with that one. And we had Chinese gangs, the Wa Ching. The Wa Ching is watching you, boy. And they were the arch rival of the Joe Boys."

"Of course in southern California you have the Bloods and the Crips,"

"And the Flying Dragons."

"And the Feudal Warlords"

"Not to mention the Gangs of New York."

"Hell's Angels, Banditos, the Gypsy Jokers, Pagans,"

"In San Antonio, they had the Warlocks and the Banditos."

"Well there are white gangs too. The Aryan Brotherhood.

And all those right wing Nazi militias and gun clubs."

"And the hunters."

"And the cops, cops in every town."

"The DEA, the FBI, the CIA. And the U.S. Marshals.

"What if these started forming gangs. You know they would. State police, county police, city police, housing police, transit police, Indian reservation police, prison guards, and private security companies."

"And there would be armed Moslem terrorist cells, or enclaves of the Jewish Defense League, the Fruit of Islam, the Ku Klux Klan! There are the Skinheads, neo-Nazis, the Christian Identity movement, armed religious cults, the Black Guerrilla Family."

"Oohh that would be Far Out! The Ku Klux Klan against a united Bloods and Crips. It would be some kind of awful. It would be like Rawanda and Bosnia."

"In the worst case scenario organized government will vanish entirely, and the fighting will degenerate into Mad Max — roving bands battling each other for access to food, loot, liquor, rape, and sheer survival. This is not the same as organized guerrilla warfare — they at least would have some ideology. This would be total, society-destroying anarchy — where children have guns. It has been going on in Africa for what seems like ages, like Rwanda, Somalia, and Liberia where it has even included human sacrifice and ritual killing."

"Yes, that old, old time religion."

"In Africa and Afganistan and Cambodia and all those Golden Triangle countries you have warlords, where the gangs take over the military, and BECOME the military. They are used to this."

"The scramble for food and water will lead to genocide and ethnic cleansing after the apocalypse."

Wild Bill: It doesn't even take an apocalypse. Look what

we did to the American Indians out of sheer greed."

"Yeah, they really got savaged by the white people."

"America is saturated with armed organizations that will degenerate into the ethnic armies come the apocalypse. It might be the Blacks against the Mexicans and the Whites all against each other. It is a big question: Can the bond of an armed service transcend ethnic identity."

"I don't even have a gun, do you?"

"No, not really. I have a little 22 derringer," said Wild Bill.

"Oh. That's cool."

"I keep it in the glove box."

"Damn, I'm going to be out gunned!"

Wild Bill knew a lot about the Civil War and the Viet Nam War. Several of his friends in the Texas Gang had served in the green hell of violent guerilla jungle warfare of Nam. These old friends felt safe sharing their deepest horrors about it to him. That became a part of his book *Tales of the Texas Gang*. Wild Bill set them in 1865 after the Civil War, out in the old Wild West. In current time the Vets all think there will be some kind of an apocalypse and America will break down into race war.

I started a new job working for Tao Ono, a construction company. We were working on a big condo conversion project down on Wood Creek. It was alright. Regular money. They sometimes went to a deli and bought us lunch. We had some women on the crew. One in particular I found very disturbing. She was wild, attractive. She had dark skin, was some kind of Mestizo mix of Indian and Mexican. She was an accomplished journeyman union carpenter. But what drove us wild was that she would work topless on the hot sunny days, whenever she could, just like the guys. She could do anything a man could do, except maybe throw around big sheets of 4x8 plywood. She really knew her stuff and led a crew. Some women on

that crew too. She had dark curly hair, and flashing Apache eyes. She was a lesbian. She had a little house in Clarksville that she bought and renovated herself. There she was — on the roof above, topless. She was their leader, this dark-eyed dusky woman, working like a demon in shorts and carpenter's leather apron. We were very advanced in Austin.

The tribe of the Noble Savage survived. The survivors learned as they went along. They were lucky. They all wore utility belts with multipurpose tools, like claw hammers, hunting knives, hatchets with hammer heads, knives for opening cans or whittling small tools out of wood and the hammers for scavenging through loose debris. Or for reinforcing shelter. They all had managed to acquire side arms.

At first they searched for canned food wherever they could. Hope was, to make their way to some place that was safe, to be able to grow crops, for canned food would not last long. If the can is intact, not dented or bulging, it is edible.

The small tribe was ideal for the basic defense strategy of evade and observe. They traveled light and had scouts and dogs on the periphery. Staying hidden while observing threats is the best way to determine a course of action. They traveled at night, across country, avoiding the roads as much as possible.

For most of the first year they had to adapt to nuclear winter because the fallout blotted out the sun. They established shelters that are both out of sight and easy to defend. Without the security of a government or police force, a visible campfire is almost as dangerous as not having a campfire at all.

They became good at camouflage. They wore army gear, fatigues, that would blend in with the surroundings. Even the most out of the way, well-hidden shelters can use more security. Use your surroundings to hide your home away from desperado eyes of the many dangerous plundering gangs.

There was no longer any big culture, big economy, big

utilities, big supply, big education. Everything was local, immediate, and the world was VERY dangerous. It is clear that the apocalypse was caused by humans. It devastated the world. Some say it was the computers achieving a kind of malevolent sentience and targeting humanity. At the same time — it might have even been caused by it — there was global pandemic. This happened when biological weapons under development in the labs of zealot scientists got out of the lab and headed for what was left of town. One group, given permission for a fatwah by clerics of a middle eastern caliphate had plans involving small pox. The other group, a Japanese splinter religion was working on a contagious air borne nerve virus for Asia. This pandemic alone wiped out more than half of the survivors from the nuclear war.

At the beginning, the fallout terribly exacerbated the global warming trend. The damage to the ecology was severe though not as bad as some had predicted. Some of the ecology did shut down for a while. Some key players like pollinating agents went extinct. Food production all but ceased in many areas. The hungry were soon fighting to the death over scraps of garbage. Millions starved, and millions more died from infectious diseases. Food riots led to artillery blasting cities to flaming wastelands infested with psychotic snipers. Packs of feral dogs roamed the city and tore at charred corpses. People went into survival mode. Immediately they were dealing with no food, no water, no electricity, no gas, no energy, no manufacturing. Groups banded together to take and hold whatever resources they could. They had a few hunting rifles and a limited amount of ammunition. It was kill or be killed. The Noble Savage grew up with the survivors, the isolated mountain people in tunnels and mines, who managed to survive and bring progeny into the world so there would be a next generation.

The man-made apocalypse of war, famine, pestilence

and death seemed to have no end. Life was lived on the run in survival mode, constantly in search of food and water and shelter. This made people more like animals, banded together to engage in looting what little was left. Experienced thugs and armed ex-cons had the necessary mindset to survive. First by making weapons out of the rubble — shanks and flame throwers, and bombs. They quickly looted armories and pawn shops and sporting goods stores for firearms. The people who did survive, had to go mad and back it up with a mishmash of weaponry. Now, there was always dread and fear, for the apocalypse had truly been visited upon them.

I did mention to Wild Bill that what we thought we knew about radiation fallout was pretty much incorrect. It falls off pretty fast, if you can stay out of it for a week you'll be all right. What we thought about the invisible, odorless radiation poisoning was more like an anatomy of our worst shadow fear. My generation was born and raised in the Cold War, and had internalized Nuclear Dread into a hopeless nihilism which was too much to bear, so it got transferred into greed, a mad grasp of resources and conspicuous consumption — for tomorrow all this may be no more. But this desperate attitude began changing with the anti-war, anti-nuke, ecological, counter-cultural revolution.

The previous part of this story is what life was like before the Noble Savage became the leader of the tribe. In *Tales of the Wild Seed Women* it was 15 years later and the Natural World had reasserted its splendor with a vengeance. And the humans had managed to get out of scarcity survival mode into a competitive / cooperative survival mode. The world was a jagged heterogemony, each city a polis, an unallied state. Farming got re-established after people moved to river valleys and started banding together and growing food.

These new cultures clashed. The Noble Savage was the leader of a small tribe because of his physical prowess in fighting and hunting. The tribe of the Wild Seed Women were about 30: 10 men, 10 women and 10 children. They were hunter gatherers, not farmers, unless they happened to be staying in a place for a season.

They lived by killing animals and catching fish. They occasionally found stores of canned goods. If the can is intact, not dented or bulging, it is edible. The art of pickling preserves persevered. Mason jars were prized. The tribe of the Noble Savage and the Wild Seed Women would come to a place and they would confer in council to decide if it was a good place, if the spirits of place were calling them to stay there and they would stay on and live there. To get some relief from the relentless search for energy and security was a blessing. Now it became necessary to defend against brigand gangs and worse. The tribe established shelters that are both out of sight and easy to defend. When you found a place that was good you tried to stay on, but it was also wise to abandon it in the face of overwhelming numbers of marauders and cannibalistic humanoids. There is only so much one can do even if you have a fighter like the Noble Savage on your side. Violence became commonplace. The men and women were experienced at it. They had to survive endless close calls just getting away from the many evil armies terrorizing the land.

If you read Cabesa de Vaca's autobiographical account of wandering around in Texas after he came here as a caballero in an armada of friars which got decimated by weather and bad luck, when they came to save the souls of the savages in the Mondo Neuvo, you will get an idea of the general hostile relationships among the tribal people in post-apocalyptic Texas. In the Cabeza de Vaca account you will also see the story of a man who was such a staunch believer in his own

spiritual practice that he carried that sacred time with him wherever he went, and always, no matter what kind of awful hell he was going through, managed to say his Vespers at the appropriate hour in the afternoon.

Another, more abstract insight into these post-apocalyptic times would be found in Levi-Stauss, an anthropologist who had a profound life-changing experience from doing field work living with aboriginals. He showed how the social archetypes of kinship and family that go into how thought is organized, indeed how cognitive evolution progresses, was based on algebraic structures that were invariant across cultures. His theory of structuralism had a profound influence on how we understand our culture. He was helped in this by applying principles of group theory to the primitives of mind — the mores and taboos. For myself, as an artist learning to live on nothing, it was helpful to read structuralism, to practice the science of the concrete, to try to get into the *pensé sauvage*. Being with a group saves your life.

For a while the Noble Savage and the Wild Seed Women had a place with a clear stream of water and little pools in the rocks where there were plenty of fish. They set up fish traps and snares for deer and other animals. The world of the future had an overabundance of feral pigs. The Tribe at times had to negotiate with the locals if possible. Barter was useful and efficient. Sometimes tribes amalgamated. Often they did not reach accord, often there was need for sheer physical violence and kidnapping and hostage exchange.

Under the tutelage of the Noble Savage, the men became heroic at defending the tribe against marauders. He had a sense of fair play and tried not to take life unnecessarily — never for sport. He did not tolerate those who did, and dispatched them summarily for they had transgressed the code. The tribesmen spent a good deal of their time making

better hunting weapons, sharpening knives, keeping weaponry ready. The women were the mothers that had raised the children and as always the children they raised from helpless, loved and respected them for it. The males were *compadres* at arms to help with the fighting and gorging and spiritual quests, but these men had to keep themselves in a state of attention for the women of the tribe were wild. For even though these men were more powerful physically than them, the women had much greater social networking skills. These women of the tribe were plugged-in. Into intuitive networks that shared connected purpose with the living earth's raging will to live — the "mind" of the ecology. These were the Wild Seed Women.

These women were connoisseurs of drama and pleasure, extraordinarily adept at finding the best, most comfortable places to be. They got direction in their dreams. They were able to consciously use lucid dreaming — remote viewing to cross the boundary of time and distance. They of course used drugs, ritual chanting and dancing, and orgies to open the doors of perception and show the way. The Wild Seed Women generally kept the males in a cooled-out love frenzy.

The tribesmen had a great rapport with animals and birds; they received images from the animals telepathically. They lived out the true meaning of totem. The women might be witches or sacred vestal prostitutes of The Temple of the Goddess. The women befriended animals — wolves, pumas, eagles, mice, and could speak to them in their own tongues through the tutelage of potent potions and incantatory rituals. There was lots of naked dancing and powerful transformative magical sex. In addition to being physically powerful and naturally healthy, the women in Wild Bill's world are vain and enjoy the flirting of wills with strong men. In the Wild Seed Women the ancient game of lust between the sexes is an art. They employed all the techniques: revealing wardrobe,

flattering words, alluring glances and beguiling scents, and one Wild Bill liked, flashing cheeks. Not to mention they could flat outrun any man over a long distance. The women had supple strong backs and great dexterity. And they kept the upper hand by a goddess-inspired polyandry in which the young people exchanged love in a rotation, the women took turns being the wife of each of the males, rotating on a nightly basis.

The women are wild, and know in their instinct that it is their biological duty to capture the best possible seed so that future generations will have the best chance to exist.

The Noble Savage was there to conduct the Wild Seed Women to their heart's design, though on the surface of tribal life, he had to give the impression of being in charge. He was the chief who led this nomadic group of soldiers-at-arms and their wild women and children on a constant search for food and shelter in a hostile, warring, though fertile landscape.

A big part of the story seemed to be the acquisition of weapons. There was a lot of rubble and it became their resource. People had nothing to loose and were capable of anything. One had to save the scarce bullets until they found more. The Noble Savage was a good strategist, as well as a hunter and fighter. Weapons from other epochs of time, scavenged from museums and pawn shops and armories, were often part of an encounter.

For example one tale of how his little band of desert, sand-Arab, hippie-ranger, types fought a horse-mounted cavalry. This tale had them, as they often were, starving and struggling on a long trek in the American wasteland. Theirs is a story of diaspora. But at least now they didn't have to worry about surveillance from helicopters. Very little gas or oil refinery work was going on. To avoid being rounded up into some kind of concentration camp, the tribe set about drawing the horsemen into a blind arroyo, the egress of which they block. Then the tribesmen proceed to slay the pony soldiers from

above like you would a herd of mastodons with spears and leveraged boulders against pistols and rifle.

In another tale the Noble Savage goes to rescue some of the women of his tribe that have been captured by a band of Mexican white-slavers. The tribesmen used the way their amorphous spirit, connected to totem animals, is able to communicate with real animals. The rescuers convey the intricacies of escape through intuitive horse-whispering, which travels by word of mouth through the herd to the women being held, trussed in a stockade. The slavers are slashed while at campfire, then drawn out into the bushes where the tribesmen are buried and they leap up and use knives against the armed human-traffic cartel.

The struggle for survival got savage indeed, and your only hope was trying to have a good crew around you. Also to know the terrain—pit traps and other setups were everywhere. The Noble Savage has to rely on his animal cunning and physical prowess to protect himself and the Wild Seed Women of the harem. Though it was not all struggle. There were many good times. It only took 20 years for the earth to rebuild the natural environment, after the human load had been substantially lessened. Many places beyond the cities were untouched. Animals like elephants and others that had escaped from the zoo, got together with newly liberated farm animals. Wild pigs were especially prolific. You'd see them at the ruined malls, stomping around with impunity.

It took Wild Bill a couple of three weeks to get to this point in the story.

On April Fool's Day Laura and her best girl friend Wendy came over to get me to go with them into Pease Park. They started telling us about their adventures on the camping trip. Wendy had finally mastered the cartwheel, in the mud, while on acid. They talked about the compassion of

mother earth. And general feelings of good friends. They got serious for a moment because they knew we could listen to them seriously.

Then Wendy changed the tone and sang this to the tune of *Camp Town Races*.

> Cramptown ladies sing this song: Midol Midol
> Lasts 4 days of a Lunar month long
> Oh, Midol days.
> Gonna bleed all night,
> Go-nna bleed all day.
> I'm tired I'm bitchy and I can't half see
> So you better stay of my way.

Wendy really impressed me with her astute analysis of her dad's emotional life. It was amazingly sophisticated, well beyond my own capabilities. I must admit it made me feel a little psychologically retarded.

Wendy, Laura and I went to the park and nothing much was happening, because it was raining. Then we went, in Wendy's car down to Barton's Springs to go swimming in the rain for it was a warm rain on a spring day that is humid and hot and sticky. The water in the aquifer was cold and good like a refreshing shower. The water comes from underground springs, and flows over turquoise rocks making it emerald green in the light. It is big, 3 acres. Especially nice on hot summer days, where an armada of flotillas loft on the undulated cool dark blue green water. Funny how when the girls were changing into their shorts there in the car, we started making up ribald songs: Don't show your meat in the parking lot / Boys will come and think you're hot / Beat your fists through the static and the noise / Open the window and moon the boys / Are you going public soon / Why so sad and blue?

Later, we went back to Laura's house and picked up her little sister Elizabeth, 7, pining away something terrible because she had been left home alone. So we took her over to my place and fixed up a big spaghetti dinner for all. Then we wrote up a note which we were going to attach to the kid and dump her off for her mother.

Laura and I went to the movies, with Talbot. His girlfriend Vera was supposed to come, but she didn't make it. We watched *Black Orpheus*, and *Donna Flor and her Two Husbands* in the university theatre in Batts Hall. Laura seemed to enjoy this heavy dose of mush to end up a beautiful April Fool's Day. Maybe she just went to please me. Laura and I made love back at the flat on Baylor Street, then I walked her home about 3 in the AM.

On the way we got hassled by a carload of fools.

The Noble Savage saw it as his duty to lead the tribe to places where the spirit could hold them in thrall. There were pockets of natural beauty, Shangri-la valleys where nature had prospered and returned, and it was in these places the tribe sought to dwell.

The tribe had collective dreams. In them the women had a way of finding comfortable places to be. They got direction in their dreams. There was one Tale, in which Cassandra, a warrior woman and the main wife of the Noble Savage dreamed that several women and some of the men of the tribe were swimming in an Emerald Cave Pool. There was a jaguar with them. They were deep in a green forest.

The women of the tribe were naturally strong, in addition to being beautiful. Wild Bill's descriptions were like Greek statues, their back sinewy with elegant lady muscles. He had them vaulting over fences, and even lifting vehicles off of children when extreme rage / adrenalin kicked in.

They also had pronounced intuitive abilities. They had a special relationship with Nature. The dream meant that they would be in a place of beautiful water for swimming and that they would have a real Jaguar in their tribe. The world in this bleak future, is a surprisingly lush landscape of abundance and generosity through which the survivors move. Nature has returned with a vengeance to assert itself, after wiping away much of the human load that had been the pinnacle of her progress, but was now interfering with it. Theirs is not a patriarchy world, nor a matriarchy world, but a goddess world.

There were scenes of beautiful young women with long black hair and bronze skin entering the emerald water hole, an access to the aquifer. The water was dark as was the jungle that had grown up around it. The tribe bathed in the Emerald Cave Pool. The tribesmen are swimming naked in their jungle pool. Adults and children. One of the women is the seer, over near where the water comes through the rocks. She is psychically the most alive in the tribe, her name: Cassandra. She is relaxing there. She often has a troubled, occupied look on her face, especially when she is visited by episodes of "seeing," though not today. There is little Thomas he is a hellion, about 6 , Maria's child by the Noble Savage, he is running around with another little boy Pascual the son of one of the other men of the tribe, with a woman they picked up from Mexico, a dark skinned beauty. The young women are openly fraternizing with the man with whom they are "wife" on the current rotation. They have found that this contributes to group tonicity, though it can lead to jealousy too. They work through it. It is best to be passionate about your comrades. It enhances psychic ability, which the tribe came more and more to accept, as their gift.

The Noble Savage's mother, Lilu, is still with the clan. She is not very old and has white hair and has no trouble

keeping up. She is so valuable in terms of her wisdom and inspiration and spirituality, especially with the younger women.

There are a couple of little girls too, sunning themselves on the rocks, they are beside a tall skinny guy who does a lot of the farm work, and another woman, one of their mothers, and another man, one of the girl's father.

It was good, to be in this place, the tribe can let down its guard and feel decent and clean and blessed for a while in this place. There was a beautiful teenager, named Huisa, tall and rangy with little skinny shoulder blades and long, thin, strong legs. She is shy about her tiny breasts. The men are off to the side, they are smiling and being like children again in each other's company — lean and ripped warriors, scars healing, dunking and roughhousing. They have left only two guards, for it is safe, and they have *seen* it is safe.

The people of this area organized work, went into the fields to plant crops to feed themselves and to barter. The land was organized into fiefdoms, with the warriors providing protection for corn and beans.

As a writer, Wild Bill had a gigantic gift in that he was able to recall past lives. Though he might not use the term, he was a Transmigrational Writer. In his tale, Wild Bill used this extraordinary natural talent to do his writing. He could "remember" things from an ancient past by tapping into the collective unconscious. Wild Bill was gifted with a singular ability that any writer would love to have, a kind of transmigrational imagination. He really did feel that he was able to see into his past lives and this allowed him a rather unique way to define humanity and its more simplistic needs. The easy nature and smooth dialect he used to describe the arresting beauty of the women he created and how the men who loved them held them, was powerful and graceful because in a very real sense, he was there experiencing it.

He told me: "I can be whelmed by psychedelics. I like acid, peyote/mescaline, psilocybin/mushrooms, if it is natural it is OK. Indeed, I see interesting things I might not be able to remember. But in the world of man there is some horrible shit, and in places in my history is such, man killing probably not even the worst. In writing I do this self examination and I face it. I have not necessarily been so tough. I have kept building my heroism."

Wild Bill described directly, as experienced, this dream / knowing. I wish I could render it in text the way he did.

It would be direct, something like this:

Their lives were broken and they could not uphold the image they had of the Other.

It had to be restarted all over again.

Love was in a prison of fear. It was up to the Wild Seed Women to bring it back from the brink. Our love.

Everybody was alone on the earth.

The tribesmen were no longer just people, they were more like animals. They had a stoic impassiveness, an openness and a kind of animal grace.

The tribe got direction from a kind of soul intuition. It taught them things that had been forgotten by people. It was the science of concrete, like the preparation of herbs, the building of canoes, the construction of dwellings. It gave them directions in their intuition. It used symbols and images, videos that played out in the subliminal opera below the waking mind.

One night Laura and Isa and I went for a midnight swim at the Deep Eddy pool. Me and the lithe young women easily moved through the shadows and climbed over the low cyclone fence by the main entrance. It was a full moon night.

"We'll just do a quick in and out," I said.

It was glorious to be naked with these two lovely young women. Swimming around in the cold fresh water, midnight at the Deep Eddy Pool.

The pool is huge, vast, long. It is a wader's pool. Only up to your chest at the far end opposite the steps of the shallow end. There is a separate deep water diving pool, just on the other side of the wall that separates it from the wader's pool. There is a lifeguard platform at this dividing wall between the two pools. It was there that I told them, "Pile your clothes over here near mine, here by the lifeguard tower. Then if we have to make a run for it, we can be as far away from the entrance of anybody coming, as possible. And that would give us the most time to get dressed and jump over the fence."

These young women looked like they could vault a fence with ease. "But I won't let anybody sneak up on us. I assured them. "I going to take my clothes off and wade in. You guys can leave your clothes on if you want to, but it might be uncomfortable when we walk home."

I pulled my T-shirt off, dropped trou and stepped out of boxers. I tried to act nonchalant in the buff in front of the teenage girls. In the shadows of moonlight moving across my skinny backside they watched as I headed into the shallow end, going down the stairs really slow. Though it was a hot night, the water was bracing and cold because the pool is drained every other day and refilled fresh.

Laura was soon stripped and she sat on the edge of the pool and vaulted herself all the way in, making a little splash.

Isabelle took off all her clothes too, and came around and got in by coming down the entry steps behind me.

I took steps through the water, wading through the white glow of the moon as it glimmered in the depths and

shimmered in the ripples of our wake drifting through the surface. I tried not to perturb it too much.

"Quiet," I whispered.

I was an old hand at this midnight pool escapade. I am showing you my secret, for often have I wallowed in this late night waterhole. It was practically a right of southern living that you could not deny access to your waterhole in the heat.

Me and Isa lowered ourselves down in the water so that just our heads were sticking up and we waded — breast-stroked over to Laura. We were not going to get our hair wet but Laura the porpoise, had gone all the way in.

In the moon glow underneath the giant sycamore trees at Deep Eddy pool I float and look up through the branches into the night sky, enjoying the cool water. It felt good to be breaking the law, to be trespassing, to be contributing to the delinquency of minors and also to be a caretaker of children.

The girls were whispering and giggling to each other about something.

Sinewy, skinny, girl limbs flash white and stealthily. They jack-knifed their smooth white girl heinie flashing cheeks out of the water and graceful back under in the shadows. It was a precious time. I knew young people feel things deeply and I was feeling the magic of it, the sweet youthfulness of it too.

I reminded them: "Just do the silent commando stealth stroke — the breast stroke, and the scissor kick."

But Isa splashed Laura and Laura splashed her back.

And I, acting all stern like an old schoolmarm said, "Ladies, we must comport ourselves with dignity."

And they hooted at that; but stopped the roughhouse.

The girls seemed to want to be together; I just floated. I stretched out, got my hair wet too, but wanted to keep one ear out for cops or trolls or temple elders of the business.

I was an old alligator. Or a catfish with a sweet little grin thinking about this cool teen age girls scene I was in. I swam beneath them, saw their legs and dark deltas, and little white buns in the moon light. Ahh, to wade waist deep in the shimmering waters of their youthful beauty and glory.

The midnight commandoes got out as stealthily as they got in. They pulled their dripping wet youthful bodies up the ladder entrance and sauntered barefoot over to the guard tower where their clothes were piled. The secret swimmers had not brought towels but it was a warm Texas night and they shook the water off like dogs or proto-humanoids. They were. Quickly without looking at each other (too much) we put the dry clothes on over our wet bodies and moved toward going over the fence in the easy place by the wall. The man and the two teenage girls vaulted the fence and we were out and walking again in the park and feeling refreshed in the humid summer night. Laura and I got especially close after that. We had gone into action as one to give her young girlfriend a good experience.

One night Cassandra, one of the wives of the Noble Savage, had another dream. A spirit in the ecocircuit entered her through bioinduction to teach her about how to allow the self to be entered by the spirit that lived in the cactus. (I will explain the terms ecocircuit and bioinduction below.) In the peyote cactus, nature concentrated a potent potion, a spice that facilitates space time travel. It was one thing to partake in the world of the spirit by following the One that lives in the peyote cactus apple, through this singular portal between worlds. It was another to really travel on it.

Certainly Wild Bill had read Carlos Castaneda, but his writing was always his own direct knowing. Simple and direct. I am trying to recall the gist of it.

It would be something like this:

I felt myself move like a shade. Through the forest. It was quite melted. It made things appear shifting, looming large, and crystalline in the sun. The magic net, known of old. They moved out from where they are. They moved like a shade, a shadow — their spirit caressing lightly the objects it touched. It was a touch that could feel, could read the meaning in things. They embraced the trees and the rocks with their spirit. They could become one with, and abide there — disembodied, curious, aware. Seeing.

You take the cactus like eucharist, it is a drug that facilitates space time travel by allowing you entrance into the subcortical organizations of the brain. What does it do? It helps communicate with spirits. The real. You can be the flow of energy in an ecological circuit. This allows you to act like and see through the eyes of (your totem, ally) animals in your dream. Tribesmen become aspects of totem animals, they can fly, they can instantaneously be places. Bioinduction is telepathy and when we dream, we are taking part in the bioinduction process with a being someplace else. The Noble Savage and the Wild Seed Women became anima, became their totems roaming through the countryside, keeping an eye on things and each other by remote viewing. The Noble Savage was a Lion, others were Bears, Wolves, Deer, Cats, Snakes Beavers, Eagles and Crows and Ravens.

They were seeing through their totems. I don't know how else to express it. It was like Castaneda said. A kind of cross species pre-linguistical bio-induction by dream telepathy. Though Wild Bill would never have used those words. He would just say Spirit Medicine.

On one of our walks Wild Bill said, "These free hunters would have had superhuman powers." Wild Bill used his

high-speed, moving-eye, point-of-view consciousness in his story to good effect to give a sense of how the Noble Savage was connected to the energetic atemporal in his world. Through the energy between the Wild Seed Women and the Noble Savage, Wild Bill was able to project a world beyond the current temporal and spatial vacuum, a world in which he could work out many of the issues he considered. "These men would have been physically bigger and more powerful than the men of today, and these hunters would have had intellectually superior minds, in touch trough a clairvoyant rapport with the natural world."

The *Tales* were a critique of the depersonalized robot-zombie earth we now inhabit. The Noble Savage was not one of the money grubbers; not one of those to dig in the earth, barley eaters. He was a hunter: alone, stalking, living on the land for weeks, able to run down game for miles and miles, who knew from observing the borough and dens of animals, and respected them, learned from them."

The women were drawn to the Noble Savage like they were drawn to beautiful powerful wild horses, wanting to run a hand over elegant muscled haunches, wanting to wrap their legs around his body and ride him, feel his energy in musculocutaneous rippling undulating beneath them. Let themselves be held in his great strength. The Noble Savage wanted to find the way in himself, in his genes, to pass on a future to his offspring — his proclivities, the hunter, the warrior, the man of insight and knowledge.

In Tales of the Wild Seed Women, he went into the future by going into the past. And though the future is a dystopian one, it is also a future filled with powerful feeling.

Wild Bill was expressing the heterosexual need. He also said they would have practiced anal sex too — for this was a time before birth control. He seemed to be speaking of things he knew. I would have asked him, How do you know this, but

knew he would probably tell me he had actually lived among them in those times. He had a transgenetic memory of it. Nothing like first hand experience.

In their dreams, the tribesmen entered a place of intersection, were time future was overlapped with time present. They saw themselves doing things in the dreams usually through the eyes of their totem animals, and then later they would find themselves in the day of their real life in the same scene. Thus they had a blueprint of their future's designs. Then the insight of an individual had to pass group discussion and come into their collective intuition through an élan vital. This happened when the intuition was so strong in the being of one of their members that he / she was able to convince the others of the authenticity. Sometimes the spirit visited the others too at the same time in a group dream. I am calling this the "ecoecho" spirit. This is where the ecology is echoed in a kind of inner mirror. The sensation was more like localization than seeing, the way one localized a sound heard even though not looking at it. The ecoecho behaved more like waves than space. It goes out and it reflects. It wavers. It shimmers. It suffuses. It drapes. This sense of the generosity of the ecology is felt as an intuition, a resonance between the left and right side of their bicameral minds. The Medicine gave them access to preliterate time. Theirs was a pre-literate culture, a pre-linguistic knowing, an animal knowing.

A couple could absorb this "Spirit" or "Medicine" from each other in sex. It was absorbed or ejaculated through his wick wet in it. Women were the drug of choice for the Noble Savage. In sex they truly did become the beast with two backs, moving together in a shared dream. They saw places they had been or places they had never been but toward which they were going. Here is an attempt to describe it. The

Noble Savage and one of the wild seed women are together, in front of a stone slab altar. There is a small flame on the altar.

Having sex with one of the wild seed women, he could feel himself elongate in her body, elongate out through the tunnel into the light and flow and fly into another sky.

They became like fish swimming up stream, diving and rising and plunging again. Going to that place where everything begins to collapse into a tunnel — heading toward a light seen when the top of the head lifts off. The medicine, swimming in Her body. Swimming up stream you become like the log in the fire — that bursts! Shooting sparks, sparks that float up and away from the camp fire into the dark sky, becoming the fire-flies and then the stars. As he swims into her, they became as one entity. He could see through her eyes; she could take on his power. And their cores could touch and fall into each other and become part of a bigger entity, the abundant green living planet swarming with life everywhere — Gaia of the great aspiring incubating altruism. It was in the land and the sky, the net of destiny. In the geology and the atmosphere and the social sphere. One saw the next place — a watering hole, where there was flowing water and plenty of fish. There they saw planted fields, so it meant they could rest there a good while.

The animal spirit filled them with its élan vital, it showed them what to do, how to be. The interconnected circuit, the net, that had almost been broken had re-established itself even stronger than before. So that this extreme threat of self-destruction — part of the intelligent being's makeup — would not get so empowered again. For it was the purpose of this spirit to insure that intelligent life evolved to observe the way the Universe was working through Nature.

In his tale, Wild Bill wrote in a speedy impressionistic style that conveyed directly to the reader a sense of the minds

of the people and what they were going through. I will have to resort to a more discursive prose, and try to do with concepts what he did directly. Wild Bill did not use this terminology I have developed.

For example in trying to run at the speed of his impressionistic riffs, I might blurt out the following with terms like bioinduction, cipher, thought-form. *Thinking about the holes, the tunnels of bioinduction, being able to fan out into the shadows, moving catlike at the speed of light, they met in the image. The speed of image that sourced through word and passed around us and between us (them), cleaved us so that things began separating, then dissolved, or became invisible, insubstantial, apparently apparitional. Like an image. No matter what they were going through they were together. A shadow that caressed, the world. To sojourn like that was to be a spectral cypher in a world-mind thought-form quest.*

The Wild Seed Women had psychic power, and highly developed intuitive knowledge that was transmitted to them through a direct claire-audio echo communication with a world wide-entity which was given the name — Gaia, that personified the beneficent and parental caring love of the earth. I pictured this Gaia entity as a kind of "econoosphere," a neologism of "ecology" and the "noosphere," the term given by Father Thieard de Jardin — a Jesuit whose theories of cosmogenesis and evolution were banned by the Church — to a kind of collective mind that was the sum total of knowledge both archetypal and derived. From "noe" — the Greek word for "mind" sutured onto "sphere," — for the earth has a lithosphere and a geosphere and a hydrosphere and an atmosphere and a biosphere, a microsphere and a toposphere: these concentric and interpenetrating spherical shells of more and more complex organization that the Earth evolved around itself so that a sphere of trans-generational

knowledge was developed by the intelligent life that evolved in the spheres provided so that it could exist.

I pictured this self-reflexive econoosphere "world mind" through this analogy: human beings are the ideas in an earth brain. Their moving around is like a signal in a pathway on the mind of this great complex entity that is the Earth.

Ideas : Human mind :: Being's destiny path: ecocircuit.

Allow me to digress briefly about the concept of "econoosphere." For I was impressed with how Wild Bill seemed to live this ancient world view of "Medicine." And now here was an opportunity to understand a contemporary expression of the perennial philosophy, even if I did have to translate it into a system of modern metaphors / models of circuit theory and computer / computation. In trying to grasp the grammar of "Medicine," it really suggested an avenue for exploring based on my background in electrical engineering. This analogy of circuit theory and ecological process reflects the development of the favorite metaphor of our time: the computer. Here the analogy is life as a biocomputer doing all these precise calculations and variations of parameters, myriad calculations in levels like a machine language and a compiled language and a user language. The modeling of ecology on circuits leads to a picture of nature as an ecological computational entity, just as the modeling of logic on circuits leads to a computing machine. Perhaps the term bioinduction might be better for this logic of nature organizing the interpenetration and propagation of energies around forms.

The bioinduction took place at a sub-cortical interface where the neuronal ganglia of an individual brain became in resonant synch with an external eco-circuit, flowing in the pathways of gene dispersal and life force organization held and slowly made available through archetypal memory in

the condensers and coils of the living entities of the world. I use the term ecoecho circuit, to suggest an analogy for modeling the world mind as the computation circuitry of a kind of bio-computer. Just as the components in a circuit exchange and store energy, the energy in circuit, one can think of living beings in a grounded circulation. For example, the circuit of salmon: running upstream, feeding the bears, spawning, dying, then their bodies feeding all the lives down stream. And the living fertilized spawn then going into the deep ocean, to return again. This way of thinking in circuits, was about focusing on the whole containing the individual, focusing on the integration among the parts, focusing on the observer as well as the observed.

In the *Tales of the Wild Seed Women,* the net of social security was completely taken down. As was the net of big time technical economy. Other little nets were developed. The tribe, the family, the people, were trying to rebuild the net. But the human spirit, which still shone through man was sorely buried. The earth so horribly wounded as to want to say in personification to the people, "Fine then! You don't like it here? You can go." Mother Nature is dangerous enough normally, but when she gets petulant and mad at her children, you better watch out. The Wild Seed Women were always snaking their way through dense woods, sometimes even coming to the rescue of the Noble Savage.

Ultimately I got the impression that the story was on one level an allegory about psyche. The Tale had an amazing effect of illuminating a psyche, bringing out the shadow or other aspects of the psyche, as though the male and female were archetypal characters in an ancient landscape.

Wild Bill and I did have many interesting discussions about women. Just as violence in the men's book *Tales of the*

Texas Gang was about moving the soul into a state of animal grace and paring away the veneer of society, he wanted *Tales of the Wild Seed Women* to be the women's book and he was using sex scenes to burn down the barriers, and to get at their souls.

The basic premise of the book was: Man is naturally free and woman is naturally free to be possessed by man; and that men and women would spent their time on earth working out the natural complications of this. This worked itself out in the concept of attractive female as prize for the most aggressive hunter /chief. But also that female must capture the genetic code to produce offspring that would have the best chance of survival in a dangerous world.

On a dog walk one day he explained: "Hunting groups would have had chiefs that were the strongest, biggest, most intelligent men. And the bravest. They would have to be men with a great deal of heart — in tune with nature to a greater degree than we can imagine. They would have had a big appetite. They would have liked big women, strong women, athletic women that could keep up with them and help out, in addition to being physically and psychically powerful and naturally healthy."

The book was Wild Bill's way to address feminism seriously. It expressed humorously the plight of heterosexual men of action facing the "artificially fit" world of modern consumer society. The men were somewhat bewitched by women who were adapted to be attractive. This gave the women the power to select for the best seed to start again, civilization in each cycle. The women Wild Bill admired were the ones that could see beyond the artificially fit, could appreciate the wild in a man. These were the Wild Seed Women.

Wild Bill was often fond of saying, "Women require too much attention." He explained that some females, because

they are held so dearly by their parents, grow up with the idea they are above cost, or price and that as a woman no one has the right to determine or even ask what that worth might be. Yet at the same time they were a time-sink, they required a lot of attention, training and what not. And so because they are so overvalued, a man has to figure out how much time he can devote to making himself worthy, presentable, marriageable in our society.

In his story, Wild Bill was commenting on the way he sees our current circa-late-1970s society by developing an extreme contrast: a world in which women were exchanged like chattel. This is how things were in the ancient earth and how they will be in the future earth after the apocalypse. In *Tales of the Wild Seed Women*, the women are always being sought after, sometimes captured — once by slavers, another time by a totalitarian regime, another time by a cult leader. Of course after the apocalypse the women didn't at first care for the return to this old order, which was more or less there in our times, anyway. The *Tales* played with that desire and frustration that bedevils all men and women: the great need and empathy lovers felt for each other. Wild Bill would often say, "We come from woman. We need woman."

In the world of the Tales we see that though the Wild Seed Women are free, they know the value of being bound to a superior male. And they know that they themselves are of value to those to whom they are bound, those who love them in companionship. Wars, arranged marriage, transaction of self, mythology, financial gains, prizes, and treasures all depict women as prize. Yet at the same time the Wild Seed Women were self sufficient and indeed often had to rescue the Noble Savage. Wild Bill plays against that stereotypical received idea of conventional male dominance in this. Woman as prize ratifies the idea of male dominance making it unseemly for a woman to rescue a man.

The way the band of survivors organized their loyalties around sexuality was a confrontation to the intentions of woman-hating groups that use cult measures to transform relationships between men and women. It was Wild Bill's view that modern society had turned into a kind of cargo cult of object worship that manipulated the relationships between men and women at every turn, at every moment, starting from when boys and girls got lined up into separate gym lines in grade school.

He once worked it out with me on a walk. He said: "Well, hunters didn't sacrifice virgins."

"That was done," he said, "by agricultural economies where men were more dependent on the forces of nature, the weather and other forces outside their control. Those economies grew big and had specialists like priests and kings and witch doctors — these are controlling types who required that you believe in them, so that they will exist and continue to get things over on you. It's the same today. Though it is true there were predatory marauding armies out there trying to get into the granary, that was an enemy you could fight. Thus the warrior class."

He shook his head attempting to comprehend the incredulous. "These agrarian economies were willing to sacrifice one of their most precious beings — an innocent young virgin girl — for the good of all, to somehow get some kind of control over the unknown.

"Now a hunting group, would never have been able to get itself that big, so as to be that out of touch with its own members. Each member would have been too necessary, too important, to be without. No one would have been so expendable as to be sacrificed in a ceremonial homicide."

In Wild Bill's fiction, the hunter-gatherer tribe often contend that man is naturally free. And that woman is naturally free to be possessed by man. But even for them, the

issues are more complex than this simple formulation would suggest. Rather, it turns out that there is no higher person, nor one more respected, than the free wild seed woman. Yet also in this post-apocalyptic mythology, the ancient war between men and women is evident: the men have more physical power; the women have beauty, adaptability, fertility and a more highly developed empathy.

Wild Bill and I did wonder, what will be the role of man in the future when science learns how to make test tube babies. We did remark on the situation with genetically altered salmon, that the males were so much more attractive to the female salmon, that they would only want to mate with these artificially enhanced males. And then their offspring would be sterile and that would be the end of the salmon. And the many lifeforms that depend on them. It is good to have more than one generation in the dating pool

Recently we've been discussing human emotions and how it applies to Mastery and Slavery. We were trying to figure out why in our time, Love felt like slavery most of the time. He was trying to find a way out of that. Or at least understand it. And his writing worked out that question.

The Noble Savage is situated in a tribe. He has various females, a harem and he often has to intervene to keep the women from fighting over him and beating each other up. Wild Bill was interested in passion and violence. He was a man of action. The notion of 'prize' is quite ingrained in the relations between men and women. It comes down to obsession and possession. Given that in the typical hunter / gatherer tribal life, every person in the group is integral and that there evolved a natural celebration of the intelligence and beauty of the human female — a form of life so remarkable, fascinating, exciting, and desirable to the male — it was natural that the hunter cannot be content with anything less, than its possession. Beyond that, the Noble Savage

uses his natural powers of being attuned to nature, to wage war and protect himself and those in his tribe, against those forces that would enslave them.

On our walks Wild Bill and I would be talking about wetbacks who lived free on the land and fished in the river like they were the tribals. We regarded businessmen as if they were aliens. In our world — so hunkered down as we were, beneath the radar of power and money, it did seem like we were being enslaved by various alien races (controlling conglomerates, cartels, legal systems, mafia, gangs, police, governments, landlords, bosses, the clerics and the clerks.) They were gaining ascendancy. And they will try to take over the world after the Big Change. It is up to all free peoples to resist.

The Noble Savage lives in a cave on a mountain looking out over a forest valley. It is somewhere near the border with Mexico, remote, wild, in the impenetrable mountains of the Sierra Occidentals. He has two wives and concubines. He does not have any neighbors coming over, for they are respectful and keep their distance from this man imbued with natural wildness, for he walks the land in quiet cunning, stealing through the shadows, entering their domain unseen, taking their leavings, capable of swift action, and defiant.

The Noble Savage stalks the women in various creative scenarios that they game play, in the countryside where they live, and downtown in the city nearby — when they want to go outside their comfort zone. He sets them loose on an errand — Cassandra likes to go to the gym and an aerobics class, Maria likes to shop, and Theresa likes to visit book stores and automobile show rooms. Sometimes he'll send them off to a beach resort, in an exotic equatorial clime, then come and find them. And "rescue" them, from any trouble they might have gotten themselves into.

Some Correspondence with Dick Ache

> May 1, 1979
> To: Mr. Walker Underwood
> 1100 Baylor St. / Apt. D
> Austin, Tx.
>
> Dear Mr. Underwood,
>
> Owners of the unit that you now occupy has arranged to hire a contractor for extensive rehabilitative work. The rehabilitation in connection with the property is conversion to office space. For this reason, we need possession of the unit you occupy.
>
> This is, therefore, your notice of 30 days that we need to have possession of the unit. Please vacate this unit by May 31, 1979
>
> > Very truly yours,
> > Richard Ache
> > T.E. WILEY COMPANY

Damn, that was short and sweet. Went out to find everybody in the five houses had been evicted. We didn't see it coming at all. The denizens started walking around and asking each other what could be done about it.

Our first idea was not to give the bastards any more of our money. Then next to take them to court and see if we could get some extensions. But it was sad and disruptive. Mandy and Robert had little Tristan in school. We started gathering materials for our day in court.

I wrote back:

Walker Underwood
1100 D Baylor
Austin, Tx.
5/6/79

Dear Dick Ache

I have received the notice to vacate the premises. Darn, just when I was starting to get this place into some kind of tolerable order. I had taken up the standing offer of T.E. Wiley and Co. that the costs of materials purchased for repairs to a unit could be deducted from the rent. I have enclosed receipts for materials purchased by me to make some necessary repairs in the plumbing.

1) washers and valve stems for the lower and upper kitchen faucets.

2) faucet handles

3) plywood to replace rotten sheet rock and plug holes in the kitchen walls.

4) studs and plywood to build a new kitchen sink stand cabinet as the old metal one was rotten away.

5) plywood to patch the kitchen floor where leaking plumbing had caused it to rot away.

6) parts for plumbing in the toilet mechanism.

This comes to a total of over $40. For the rest, to make up the balance of $85, which I was informed in a letter, would be the new increased rent due 5/9/79, I will apply the 45 dollar property deposit, being held by T.E. Wiley & Co. for this last months rent of April. This makes an access of $85.

I would be glad to rent from T.E. Wile & Co. in the future, and hope you people can find all of us here at the Hill, some suitable new quarters, although I doubt you will. This was a beautiful place, which all of us loved.

I would be interested in offering my services as a carpenter for the extensive 'rehabilitation' project I understand is to take place here.

                                Sincerely
                                Walker Underwood

A Room of One's Own

Whooo boy, getting evicted was going to be a huge disruption. Things were going so good here for a while: sweet girls coming around, writing poems, singing songs, dancing and jiving. Good friends, good food, some art was happening. Some spiritual understanding was unfolding. Now I was to be casting about for some abode into which I could abide.

My first impulse, as always, was to do nothing and let events take care of themselves. This is a cursed aspect of the INFP character type, who preferred intuitive ruminating and was not afraid of the shock of the new. However I am getting a little old for that. I am 30, and I should be getting somewhere.

The activity I hate most in life is job seeking, next to that it is apartment hunting. Now I had to find an apartment that will no doubt cost more, needing a better job to afford it.

In my memory I revisited the places I had lived in Austin. I must have been in just about every neighborhood and scene. And my mind ranged into the lives of the people I had been roommates with, who had been part of my life. Were any of them going to be around for the long term?

Would Austin still hold me? I came here in January '70. The first place I lived was with this puking imbecile of a room mate. Couldn't hold his liquor. What a horror, every time he went on a toot, he'd puke his guts out. Then Kenny and I lived out south at 2222 B Mission Hill, a duplex, rather a quadraplex built behind the IRS. After a year there, among the Air Force base types, I moved to a house he bought by the airport. Then I lived at the Governors dorm, on the Drag, and the Crows Nest with Mort Jules Windish. I was at Lyle House for a year at the end. I spent one summer at the Triangle H on

22nd Street, where my sister once had an apartment.

Then I lived in Dianna's Place, both on Enfield and a house she bought on 29th Street at Shoal Creek Blvd. Then on to Ave B, my first single place for a while. Then I decided to live with Terry at her apartment way out on Reinli. Then later the apartment Terry and I got on Speedway. I was into food stamps and dealing weed, and doing a little tutoring at the time. We got by for a while but then she got a fellowship to the Sorbonne in Paris and a teaching job there. I moved into a boutique down on 6th St. with my friends Roux and his old lady Carlotta. I pinned fabric to pattern. It was fun and creative working with fashion. We made the Manchurian Cowboy shirt. All the coke dealers had to have one. It was a few doors down from Antone's Blues Bar. The scene on 6th St. was amazing. I met Hans Otto, a photographer who documented the scene. The boutique owner was a young coke dealer named Oliver with an even younger naive wife, a genteel southern belle, from Louisiana. Somehow matters got out of hand and he had Roux and Carlotta arrested for some kind of fraud. It was ugly, painful. Time to go. I hit the road up to Canada at that time. When I came back, I stayed with Roux down on the river. Then split for Berkeley for a while.

Some where in there I stayed at Ave H with Jim Baron upon returning from being on the road. And at Robert's place on 6th and West. Lynn at the edge of Clarksville. Then I got the farm in Manor; it got busted and I couldn't stand to live there any more. I am just flashing, not in order: this rip-off place on E 32nd, with a bunch of people from Greenbrier — we were always broke and hungry; the house on 33rd where I met Billy and John; another house with Juliette. I had to stay at Kate and John's house for a while after I got busted and was completely homeless. They were glad to see me go. I was starting to loose my charm around all this.

Although I have lived almost 10 years in Austin, I spent

one of them in Berkeley, one in Montreal. Before that I spent some 11 years in San Antone, though one in Canada, one on the road.

I wondered if there was even going to be a long term. I slipped into some long self-inquisition, activating reservations about my net worth in the world. Where had I been? What had I accomplished? Would I ever get married? Have kids? I hoped so. It is spring and I'm feeling generative. Why haven't I had any children yet? Why haven't I perpetuated my kind? Is it that I could find no woman who wanted to have my kids? Eviction will definitely precipitate intimations of mortality — heavy on me tonight.

I seem to dwell in some other plane of existence — the metaphysical plane, when I should be spending more time on the economic plane, like most people. What had happened to make me so different from those devoting their lives to getting and spending on the retail plane. These people were just being real, taking responsibility for their own welfare. But a welfare that became status — at high cost.

I was an oneironaught, an astronaut of dreams, preferring to exist on a metaphysical plane — a space entered through the doorways of dreams, of art, of love, or of drugs. I have always lived more on the psychic plane, have felt out of place on the financial plane. Having never had money, I live without it. Though I worry. I have a fairly good formal education. I can read and understand mathematics and circuits. It is quite a thrill to feel the abstract moving through the real.

These spring days I am feeling very feminine, with masculine parts: I am a lover. It has been my way of finding a path on the metaphysical plane. A woman's sex is the doorway to the soul. So are the eyes, the mouth. All my life I have sought beauty. Poetry made the hair on the back of my neck stand up, that's how I knew it was poetry. I devoted my self to being in this beauty. Beauty is not just some silly lithe

lolita girl in a halter top. Nor is it the beauty of the Cocteau version of Beauty and the Beast, although that was beautiful. I am thinking more of mathematical beauty that has no beginning and no end. A beauty in the mind. A beauty like an alchemical transformation, of the soul rising up and getting transmuted into a more transcendent knowing. Or of the miracles of the saints. A beauty like how youth is beautiful — but that will always be youthful, a beauty that is here and now. A beauty like chance, and knowing how to change it into good fortune.

I first encountered beauty in youthful impressions of paradise: I was a fervent Catholic boy, going to ensure I got safe passage into this blue and green paradise beyond light by living out Pascal's wager and becoming a priest. I was an acolyte at the Latin Mass in the second grade. I was on some kind of religious high. Then I found jazz and beatniks and hippies. And I found the high psychedelic music of the literary imagination in Joyce and Miller. I made art a spiritual practice. All this world will be gone too soon from me, and while I am living, I must query it. Make it slow down, and somehow make it divulge its beauty and wisdom in a seine of words, as it drifts through a sieve of theory.

I have thus remained constant to writing— in spite of the dubious reception, and the guilt-inducing teasing and subtle rebuke from family, elders, friends. One in particular, my old grandfather, trying to be emphatic, pontificated this epitaph: "There is a race of men who quiver and shake, and every move they make, is a new mistake." It has stuck in my mind like some bad song. I wished there was some kind of mental air-freshener — I'd like to open up a spray can of Meme-Begone. The perfect antidote might be the anti-song. I thought to turn that bad meme into a punk anthem. Punk lyrics are cool. You get some kind of repetition going about what makes you mad as hell and you consider it and consider

it, and its ramifications. You chant it and rant it, until you see how cant it is.

> Chorus: Will you confront this obscene meme? /
> The one that insults you with its stream.
> The one that seeks to keep you mean
> and displace you from your dream.
> There is a race of men / and every step they take /
> is a new mistake.

> How can you work with that? Makes you wonder if you are one of them? / to wind up in the pen / or behind a desk or under the wheel — ground up by the chore / that keeps you striving for ever more, / for what you thought you were supposed to be here for.
> Don't you see that it's the machine?
> You must confront that mean machine,
> behind the eyes of those that would be wise,
> but all it does is deny.
> But do not act in haste unless your effort go to waste
> "There is a race of men and every step they take /
> is a new mistake."
> Fuck that. I'm going to make my own way.

> Chorus: Will you confront this obscene meme? /
> The one that insults you with its stream.
> The one that seeks to keep you mean
> and displace you from your dream.
> There is a race of men / and every step they take /
> is a new mistake.
> Just let it go and let yourself be driftin' into the flow /
> for you can feel all you need to know.
> Do not defend this suffocating trend.
> You don't even need to deny it, for it is nothing but words
> caught in the riot of the old brain.

Punk could be OK. Maybe these young people are on to something. We were encouraged by the antics of Isa. (Our Weesa. She is so divine.)

My father stayed with the same job 30 years, day in and day out; he was the plant manager. He was able to cope with it all by getting drunk almost every night for over 20 years. He was no amateur drunk. He could really put it away.

Beauty is a romantic ideal, and a Taoist idea too, in the sense of being wild and scientific and mystical. I have felt beautiful myself and known it. And been touched by it. You don't have to be beautiful to know beauty.

I was using the term "time flower" earlier. What do I mean with this concept? I mean the manifold way a flower presents itself in space is like the way a person binds time — in petal-like sheaths of surfaces, that enfold the person like barriers to protect against and use the transparent energies. These sheathes are like memory, like the pages of a book, or like a field, or like nets flung out. A person — a personality, has protective barriers or filters that are pulled up and relaxed down, folded closed or opened like the way a flower extends its petals. The extended you. For now, I'll just give you an image of this time flower: it is your group, the people in your tribe. The people around you who can help you get by. The Friends. They themselves often represent uncles and aunts and sisters and good teachers you have had in the past. Kinship has implications: Freud and Jung and Levi-Strauss explore this. Another simple representation of the time flower would be: The Inner Child, and the Outer Masque. Inner and outer articulating gestural movements among the several dimensions of a memory database would be another instantiation of it.

Later in this book we'll model the "time flower" as the simplest 4-dimensional object unfolding in the dimensions of space and time: the tesseract. For the moment, imagine it as a cube within a cube for the sake of simplicity. I used the word 'flower,' to suggest a more flowing manifold, an organic

subjective sense to this attempt to have a place to go in your mind — a room of one's own, in contrast to how the tesseract is usually depicted as a mathematical object, a cube within a cube. That can be a bit abstract. I want to reify its crystalline symmetry, its mechanical / analogical / optical modeling into objective existence. Allow me to just say that there are moments of liberation to be had from learning to see the unfolding of your life in the time dimension. Of course many have said this.

I am developing the theory of the tesseract as a room of one's own. A tesseract is the simplest closed 4-dimensional object. It is represented visually in the flat 2-space of the page by a cube within a cube. At any given corner of the inner cube you have the intersection of the three dimension of space: length, breadth and height, and then in addition, coming out of this perpendicular to the other three, is a 4th dimension of time. Tesserae — four rays coming out of each point. The tesseract is the four-dimensional analog of the cube. The tesseract is to the cube as the cube is to the square. It is an abstraction used in the world of ideas.

At the time, the amphictionic theatre in Berkeley gave me a huge intuition into the idea of tesseract as model for theatre. I wrote an essay about it, but that essay is lost now in all the moving.[1] Basically, in moments of stress, you want to feel yourself inhabiting a bigger space. Not just the little space you are hammered, socked, screwed into at the moment, but feel you are part of a bigger space with more room to be, and room to breathe. Perhaps your art can even help other people be in that alternative space with you. The model is an intuition both scientific and mystical, because it deals with a gestalt that shimmers in and out of the light of reality.

You must get this intuition, perhaps even an epiphany

---

[1] I have constructed an expanded version of *The Tesseract Theatre* in the appendix of this book.

from your own time. The ability to shift into the dimension of your own real self will have as strong an influence upon your world, as your world will in supporting it. Maybe it is breaking the chain of conditional probabilities by getting into pure states.

Tesseract Ode

It was becoming, once again, out of reach
to have a room of one's own.
It worried me.
I felt so confined . . .
choices blocked — so in a bind!
I wanted to furnish my mind
and move into it.
I started to theorize. . .
What if the points of view in the psyche were projections of a larger, higher dimensional reality outside.
If one could have more easy access to that space "outside"
maybe one wouldn't feel so trapped all the time,
— 4 walls closing in.
On paper, in my mind — the tesseract glowed
like a neon sign out there in the landscape of the free:
a cube within a cube — shifting shape in the blink.
I trained myself to think of the way to see it,
in the abstract imagination: a crystal within a crystal,
the POV projecting from within and through and onto
the faces of its facets.

At the center of the outside cube is an inner cube
—a cube within a cube, like the universe around your head, similar, alike — within and without.
Inside, a child is arranging art in a room
— this enormous room we are given to inhabit,
a mnemonic theatre full of craven images
rising to become signs.
Remember?
We are all stuck here like fish in an aquarium

```
|           } . . * >                    |
|             } . > • >       } .| • >   |
|             } . | • > < • | . {        |
```

darting around, attacking, feinting, threatening
—imprisoned in some too small VOLUME,
behind clear glass, in a forest of fake rocks,
confined and defending our little status.
    Little ones dart at little ones. Big ones with long dangling
tentacles dart at their species.
    Other denizens outside the box don't exist,
there is nothing but the aquarium. Yet
Outside there is the whole larger world with its denizens.
To the fish we would seem paranormal — perpendicular to
their world, coexisting in a parallel world — outside the tank.
That's exactly what I am talking about,
A gem, whose facets are movie screens.
These kinds of screens we see there be:
screens for looking through, 1st person sees what's going on;
screens upon which we are seen from the outside,
  — things happening to us in a movie we are in;
and multiple screens, like the all-knowing eye
that floats in the sky and sees everything, even into other minds
or the time perspective of the hero with a thousand faces
living out the story.
    Yes, to be able to travel in your mind by shifting through
the 4 points of view — coming out of any corner:
1st person, the I — seeing through the I's eyes;
2nd person the you — talking to the reader or yourself;
    3rd person limited, the he, the she, the it, an individual from
the outside, seen from a camera following the subject around;
    3rd omniscient, the all-seeing eye, able to follow around and
zoom into the mind of any character in the field.
    Now I realize that 4D objects do not actually exist in the
physical world, but just thinking about it opened my imagination.
It was soothing like having a private room of one's own to crawl
back into.
    To go there you have to discipline the imagination with the
abstract. To get there we have to travel on an analogy.
    This goes back to Plato who recognized the built-in progres-

sion of dimensions: a point in space gets slid across the horizontal to make a line.

The line gets slid vertical to make a plane — a screen in front of us.

The plane gets slid towards us, and behind us sweeping out space. And we are in a cube room.

(But what is outside the room?)

You can't see it?

No one can,

for we are evolved in our perceptual space of three.

Dali depicted Christ being crucified on a tesseract folded out to be a cross of cubes.

Let us go there, where the cube gets slid in a 4th dimension, perpendicular to the other three — into time, the paranormal dimension.

And then, I think it was when I realized,

that shadows were made when objects of a lower dimension blocked the light (vectors) of a higher dimension,

that it might be the POV room I wanted in my meditation.

Yes it was there I think.

A line or a plane can cast a shadow in space.

But it takes a cube to cast a "shadow" in the 4th dimension.

It's like when Jung explains how we have to look to our shadow side in order to integrate, because that is the place where we repress our fears, and where they condense and amplify and attach themselves to other phenomena.

One tries to have a creative life and use the forms of art

to hide and reveal one's love, one's desire — to one's self.

One is inhabiting a tesseract room to be able to travel toward and away from one's fears, isn't one?

It's like in sci fi, where the Tesseract, since it has such natural, easy access to time, can teleport beings, by lifting them out of one dimension and putting them down in another, alongside of, beside, near, resembling, beyond, apart from, and abnormal.

The paranormal in the tesseract!

Parapsychologists would like it.

This teleporting is the explanation

of out of body experiences.

And the rest of the normal paranormal stuff too.
It gives us hope that we will someday, some way
communicate with other beings like us in other worlds
— man overboard becomes man everywhere.
The man with the X-ray eyes
who starts to see through things—
the solid skeleton inside a body,
or like in CAT scans and MRI, imaging slices
stacked in shape shifting sheaths.

Time is the confounding variable.
And because it is associated with an end, it is always lurking.
(Sorry to bring it up.)
But we have the abstract imagination to entertain our
complexity before we have to give it up.
So let time sprout,
out of every corner of the inner cube,
like a bouquet of rays, — the tesserae,
generating an outer cube, an invisible proscenium.

This generating is what gives it away as a group: the series
of dimensional analogies is expressing organization in terms of
generative actions that structure a space.
But even though this is an amazing discovery, don't forget it
is just a projection at a cost to perception.
A Tesseract is not necessarily a stable figure like a cube or
a square. The image can collapse in on itself and expose its
various surfaces through movement of the cells.
This novel borrows from that concept as the flow of the
story happens in stages as opposed to a linear story.

Laura had a tiff with her ma, and came over, and hung out. I was glad for the distraction. We went for a walk. I let her talk, I didn't share my own misgivings about the future. She was going to be graduating from high school and leaving her mother's house, and couldn't wait to get out. To get a job, and an apartment. Perhaps we could continue to be together.

Austin Deep In the Wild Heart

Laura and Isa were disco dancing in the kitchen of our Baylor Street flat. Laura playing the male and doing these vampy moves, running her hand down the side of Isa who has her arm outstretched in a John Travolta pose.

Laura, in tight jeans plays the Male, and she pulls Isa to her.

Laura: Mmmm baby let me feel some of that. I love the way you feel, so real.

Isa: I'm a Riot girl, feel so holy to be alive.

Isa doing some fancy footwork basketball step. She wore hose, up to the top of her legs, and her pleated mini skirt just barely down to the top of the hose.

Yes. Now that would get her a lot of attention.

Another time the two Texas girls are buck-dancing in the kitchen. Isa is so outrageous, flouncing around. But she was so cute! Wielding this on-setting sexuality. Sitting with her legs open, lifting her dress above here head. She's got big high-heel cowboy boots on, with a short skirt. I'm starting to really dig her, dig on them both. Isa is a woman too though she is only 14, a very precocious 14. So close to the wild heart of life.

Isa and Laura worked out a little duet they sang with each other. It was some kind of a goof on a Gilbert and Sullivan musical, and some other romantic all punk-girl band and underground magazine issue they were involved with. They were both fun writers.

The two of them hoofing it around the kitchen. They brought so much joy into our life. It was so sweet and hard to beat and sad too. To be part of this boy and girl thing, they were making us feel so young. They smile at me. Bat their eyes. Sing:

Both:
High trees, arching over Parkway avenue
We'll get to the park, with us girls on a lark
He is like our dad, except he's a cad.

Isa:
Eye lashes drawn swept-up,
Opening to let the world in.
Trying to look so grown up,
And so fashionably thin in this dingy din,

Laura:
I have been working in clay,
Making roses, all day.
And I would like to work in concrete
But the boys don't want us to compete.

Isa:
Those mud men, what do they know
They only crow, as if they could go.

Both:
Doing yoga in the bed room.
Trying to make my heart chakra bloom.
Bearded hippies trying to become gurus,
What have you got to lose.

Isa:
Pretty legs and bare feet walking in the rainy grass
Carrying little shoes,
The velvet mouse trap of Austin has caught me
In its cheesy snap. And I've got the blues.

Laura:
Groovy old stone buildings, verandas, porches,
All the students want to drive Porches.
So many scholars without many dollars,
living in the student ghetto
and I wonder where this libretto is going to go.

Both:
Here in the Cambridge of the plains
It's the Austin way to always be led astray . . .
Holding hands in the capital rotunda,
Click out loud make it echo round.

Pretty legs and bare feet walking in the rainy grass
Carrying little shoes,
The velvet mouse trap of Austin has caught me
In its cheesy snap. And I've got the blues.

I did have to get stern with them sometimes though. Isa was a bad influence on my Laura. They could really start acting up, and I was not their dad. But I figured they had enough oppression in their lives and tried to cut them maximum slack. It happened when we went up on top of Mount Bonnell and sat looking at the city below.

Isa was wearing a dress, because she had smuntched her cunt into the bar of Laura's bicycle. "Becoming impaled," she said.

There Isa went on and on, singing lecherous and bawdy on-the-spot songs about the wide river flowing by and the bikers that were drinking beer and how much she hated her science teacher, woven in with parts about her life — how she wished she had big tits like Laura.

Isa's School Hater rant had this:

School is the pits, walking down those dark halls, it is what it must be like walking toward the light when you have died. The teachers are so oppressive they don't get it, they never did. They have been writing the same stuff on the board for so long they don't even need to lift their hand of the board, they just stick their hand into the grove and it goes along tracing out the form of before.

No wonder we have such a blank stare.

They are keeping us here to keep us off the street!

We are not learning anything useful. He makes me feel so embarrassed and suffocated. They are petty tyrants constantly picking at your work, like vultures.

I don't want to pay attention to this idiot. He is trying to steal my heart and my soul, trying to make me all worried and neurotic like they are.

Isa got into a biker rant, up on Mt. Bonnell, while we were looking at the wide landscape trying to get a little peace.

"And we've got bikers! Roaring their machines! Echoing off the rock walls, like the world was their echo. Buffo commandoes in their little leather vest uniforms with their jeans and patches. VROOM VROOM, VROOOOOMM!

Little white Venus was peeping out, up beside the setting sun. One wanted to say, How I have missed you. But it was too bombastic, to think — against the doom VROOM drumming."

Isa yelled out: "I'm a silly slut! Still in Austin!

"I can say anything I want to, they couldn't hear it. Can you? It is pretty as a picture book here. Walker and Laura sitting with me — the dream of the flight of eagles, over the old stone temple at the top of the mountain."

She got into these wild bird calls, then started making them more human, ARG! Ayyyeeei.

Later she got into doing blood curdling movie screams at the top of her voice, screaming across the blue horizon.

Shouts: Ahhhhh!!Ayyyeee.

Sound. She said that sounds real human too, doesn't it?

A shout to move like a shot! Through the bored despair choppers revving their motors, and cleaving the space with their jackhammer energy. It was as if they were trying to separate the horizon and defeat the magic of Mt. Bonnell, which had the power to merge the blue grey waters below with the light warm blue sky above.

Ah, the poetry and vulgarity that spewed out of that

child's twisted little mouth! She was boisterous and girlicious, into buffoonery and of course wit, pun and double entendre (sexual).

I felt like could fly like an eagle, by shamanic projection into totem, like the Indians of old. The ghosts of the Indians and the Harleys are at war, blasting the high blue air. We should go swimming in the cool waters, get out of this heat. "I am drunken and going into heat," she said.

"Oh I ought to go down there and throw a bottle at them." Isa said.

"You better not," I shouted at her. "Cause if you do, they are gonna break out the knives and the chains and come up her and kick our ass! You they just rape, me they probably kill. So shut up. God damn it."

And I had to take her home.

Bhrammm Bhrammmm Bhrrraaammm

Acting all contrite. She pouted: "I will not be a loud drunk any more; nor spit; no more mooning."

She acted as though she was reading commandments: "I will wobble in red high heels accessorized with an orange designer hand bag. We will not fight. I will be cordial. Do you want me to be your love puppet?"

"I need a man to hold onto," she crooned in a drunken sway, "somebody to hug." I held her hand and kissed it. She seemed to like that but seemed to be wanting more.

Another day, Talbot came over, with Buppie, the wonder dog, Boston terrier. We went up on my deck, handing the little pooch through the window. While Juliet and Peter were getting cranked up into an argument down below us, Talbot got into this spate of maniacal laughter. This got everybody laughing. He said, "Want me to throw this dog at you?"

Then we went over to work at Folk Toy. Everything looked a mess. There was stuff all around. The whole messy

scene just looked dumb. I didn't want to have any part of it. So I told Talbot I gotta go. Afraid I pissed him off. Hate to leave a good job.

I got a new job as a trim carpenter. Greg and I went up, to a job we saw in the paper and told the contractor that we wanted to be trim carpenters. He showed us around and we could see where some former disgruntled employees had driven hammer head or claw into several of the sheet rock walls. This made it necessary to cut out and replace whole sheets on the studs. We told the man that we wanted $5.50/hr. He said, "Most anybody is making out here is $5.00." We took it gladly. "O.K. You can start work after dinner." We went and bought some miter boxes, and a back saw, and a coping saw, and looked at some brochures on home improvements so that we could get it down as to what it was we were supposed to be doing. Greg was very masterful, experienced. He took charge. Though we almost blew it, when the old guy said to make up our "Punch List."

Cagey, bluffing, we both intuitively knew to refrain from asking, "What's a punch list?" Though we didn't know.

So I had me a new job, as a trim carpenter.

One morning, while I was waiting for Greg to pick me up to go to work, I wrote down these thoughts about group theory and its relationship to the human psyche.

I wanted to find tests that would show that mind was at its most fundamental, an algebraic structure, a group. If the mind was using group theory we could have tests so that it may divulge some of its secrets. I thought that already the Intuitive insight has been useful in my education. It is one of the most useful things in the study of science. The tests were like inverses, one of the group properties

The power of inverses, of using the things that go against you, this is one that comes with maturity. It is the

most elusive as it is deeply rooted in the dream world. Yet also in the real world of economic reality as well as the real metaphysical world of sex and drugs. Inverses or adversities, are everywhere. Sweet are the uses of adversity. The poet earns his estate majoring in Negative Capability, if you can get a handle on it. For the world of inverses and adversity is the world of challenge, the biological organism thrives on it. Life is cognizant of stress and builds up muscle and other tissue to support and strengthen weak places. It is the great stabilizing influence of feedback. Yoga is the yoking together of opposites: opposing muscles make balance. Yoking opposites going attentively.

As long as it is not too much. What good are tests that just serve to grind down the ego, and don't give any boon of knowledge for the adventure undergone.

Another of the group properties was Identity: the power of the self. Identity is that to which the inverses machinate, the source, the frame of reference. The self-adversity builds identity. The self can be many things, depending on the system. Like in Addition, the identity is 0; in Multiplication, the identity is 1. It is a structural property of all groups. If the mind is a group, then it has Closure (the concatenation of ideas leading to other ideas); Inverses (the dialectic of thesis and antithesis and syntheses); Identity (the self referential reality that all ideas are based in soma); and Associativity (clear flashes of insight or intuition).

I was spinning, my mind going into anything but the real — the situation I had to deal with.

I wondered if my being into pot might have contributed to my lack of ambition and failure to launch my professional life, whatever that might be. It was something to think about, something else to worry about.

I worked on a poem instead.

In the Thirtieth Year of My Age

It was the thirtieth year of my age.

And I took up pen and wrote a kind of biography,
it was a literary biography influenced by *Biographia Literaria*:
tracking the Works that had influenced me.
And it was utterly boring; I will not fob it off on the reader.

I must admit that my pot farm had been painfully busted
and I was reduced to working for my lawyer.
He managed, with the luck of a radical judge,
to keep me out of jail on a technicality
in the search and seizure laws. Blessed be the lawyers.

In the thirtieth year of my age
and without any visible means of support
I awoke with my friends in the community
of Baylor Street in Austin,
I had my life and I awoke in the sun.

And if you are in my time may you wake too.
To the day as it makes demands on time,
and to the night in which you dream.
The sensitive weed I smoked,
taught me about being myself.
And since I live underground
the smoke helps me understand the chthonic deities there.
It is amphictionic here, (it is a kind of light.)

The harmonic universe is a song
played on the strings of an infinite violin.
How I would like to play in the fields of distant worlds.
Yet I am here now, and I have the day and the night.
In the day I go forth to earn my keep,
in the night I fall down to sleep in my self and dream.

In the thirtieth year of my age,
I know what it felt like to be loved.
I had been touched by the angelic beauty of youth.
It opened my eyes and there she was, staying with me.
How blessed, how lucky I felt.

And I surged out into the day,
took long leaps across space,
like the angel that I was,
that she made me feel I was,
and I began listening in
to what surged up from the radio of my dreams.

At night I slipped into the universe
and during the day I marched to the beat of the market.
At night I slipped into a universe of dreams
and replayed the days, of the eyes-open street.

I sense myself in dreams
able to fly without a body,
and what I see in flying over my world, is my own past,
and what I don't see, I feel — out there
beyond coincidence during the day.
The world has given me my life
and the rest of the universe has given me my dreams.
The earth has given me life
and the rest of the universe — dreams.

And in my thirtieth year
I am somewhere between being a child and a man,
still part child and yet an older man,
and vouchsafed to be caretaker of this child for a while.

And in this life I have been given
there are gaps into which I get lost,
which I must cross to fulfill my destiny.

And the days of my life are paths that occurred,

that I followed;
and the nights of my life are passed
in the dreams of what might have been.

In the day I believed, what
in the night I had drifted through.
In the night I dreamed about
the things of the day I believed.

And the thirtieth year of my age,
showed me I was projected like a shadow
from a deeper sun:
the one that projects the sun we know
as an outer manifestation of itself,
generously occasioning everything that lived.
This is the sun to which I am attached,
the sun of the souls,
that we are permitted to rise back to, in dreams.

The Peter Pan Syndrome

Dear Mom,

I enjoyed your letter very much. You write like a young girl, the one you sometimes let yourself be from your memory.

What were these times like? The times of postwar Montreal. Do you recall the name, of the Brother, Brother Herman, wasn't it? Who was doing so much to get the Basilica on the other side of Mount Royale built. Didn't he die along in there. I seem to have seen a news reel about it. Almost all of Montreal turned out. One of the longest funeral processions in the history of the world.

What was Saint Catherine St. like, with the cable cars? I saw how you rode down Saint Catherine from Wood Avenue in Westmount. And wasn't there this big dance studio, rather ballroom, where everybody went dancing. You say your friends were people from CIL and students at McGill, and workers. What were they like? I found the people of Montreal to be the most cosmopolitan in the world. But what did they think about dancing, say. Did you over go to the L'Odeo down off St. Catherine, in the quartier chinoise just past the Main?

Ah, the hockey games, and all that cold weather.

I am up about other things these days. I'm trying to maintain a home, a happy love life and a job, but it seems you can't have any three things working at once. There's always one of these components of a happy life missing. Right now we got an Eviction Notice. They are moving no less than 8 households in 5 different houses, off this wonderful hill. I was really getting into this place. I had hoped I could stop all this moving around Austin for a while.

So I'm out looking for a new place. Trying to find a farm, maybe. There's a good chance that Neverland Construction Company will get the landscaping job here. And we could make the proviso that some of us can stay on as worker / caretaker for a while, at least on past the summer. I got my garden planted here. I could plant another if I got a farm. I don't know, though. Somehow it ain't right that they can move these families out just like that. Some of these people have got kids in school. They want to make lawyer's offices out of our homes. There ought to be a law, that residential property, remains residential property. This place is close to downtown. Let them put their damned offices out in the sticks where they are from, bunch of hustler, money-gloming clods.

It's tribal the way we are with each other here. I think it is important that even bachelors like me have kids around. We keep a watch on all the kids and puppies and kittens; there are lots of cats. 'Highly Unlikely' just had kittens.

We are doing a good job on the playground that Neverland Construction is building. It has a tree house, overlooking a railroad-tie fort. And there are tunnels and sandboxes connected in a many layered maze. There is a little pool, made of limestone. It has one monolithic slab for a bridge over a narrow part and is landscaped with flowers and lily pads. We are building a large barbecue pit, and a stone planter that looks like the battlement of a castle, like a chess piece. It is big enough for a kid to be able to crawl up into.

So I'm getting all of the exercise I can handle And I'm learning a lot about all kinds of construction and landscaping. I could really do some stuff if I had some capital. I'd like to get a truck and start bidding my own jobs, doing a little carpentry, stone or brick masonry, tile setting. God I'd like to be putting the energy into building my own place. When I think of the all the money wasted in these high rents. They

are getting 400 and 500 to dollars for the simplest house. A person has got to work all the time to pay that. I get bored working all the time.

I think a lot about going to Canada and buying some cheap land in New Brunswick but it is so damned cold up there. The sociologists and demographers say that the U.S. population is shifting to the twelve Southern states in the Sun Belt. They say by the 80s that half of the population will be there. Land prices are going to soar. I don't know what's going to happen. But it has already become virtually impossible for a single person like me to buy a piece of land, let alone a house.

Ahhh, heck. I'm getting tired of feeling like a squatter on this earth. I am thinking about buying an old monastery in one of those Mexican ghost towns for about 3,000 — plant an orchard and live out of the trees.

Love,
Walker,
Baron of the Trees.

Letter from Laura

Hello,
Would it be possible to extend my visit for a couple more days? I also have Patrice to stay with. Like I said I move Sunday and . . . Won't be around much anyway. Mostly after work or before. Won't eat all your food either, and if you should get tired of the distraction I've got a place to go soooo....
I'm really tired of this running around, I'm sure it will all subside as soon as It does though, which will be real soon! If you think I'm asking too much please say so. I got to know (Know what?) (Anything you want to tell me!)
I got an extension on that Poli Sci paper. Anytime this week and I end up taking all my finals. (Due to some sticky Rules!) EEEk! Won't it be nice when . . .

Further Worry About The Situation

Mandy and Robert were walking around the complex, talking to people about what they are going to do as far as refusing to move and staying put or leaving on the appointed day or what. It was decided to petition the courts for an extension.

The benign genteel way we respected each other's privacy here, yet felt community in Neverland will be no more. It is sad and awkward how we are going around and encountering each other with this. It will be the last time slackers like us will encounter each other; we all know each other very little, really.

Juliette is ever the queen.

Wild Bill was of this opinion: "There is something very much like the spoiled child, something very irresponsible about her laughter."

With the eviction, now once again, there was a sense that the whole edifice of my life was coming apart. I had to finish up some writing I was working on before I became homeless. However all this anxiety would sometimes cause my focus to flip-flop into a running state of Quasignosis which could persevere until reset.

Quasignosis? Let me explain.

It is a kind of foggy ambivalence — a deer staring into the headlights of oncoming doom. An inability to martial the forces necessary to get out there on the veldt and run down some game.

It is like hypnosis, except you are not under a spell exactly, more like in an oblivious fog of Non-Being, unable to see any good outcome. A state of over-specified knowing, quasignosis is to hypnosis as ignorance is to sublimation.

Quasignosis : Hypnosis : : Ignorance : Sublimation

Quasignosis is like the state of being, induced by the circuitous prose in the stream of consciousness novels of Virginia Woolf. (I try to get a little existential humor going.)

"Why are you anxious," I ask myself.

"Oh, it's Nothing," I tell myself. Quasignosis.

I want to play with this existential character. Perhaps in the context of conversation with an intelligent high school girl. I was trying to honor Laura's more advanced position in life getting through high school. Not trying to be parental.

One must be vigilant and not succumb to the mental paralysis of Quasignosis.

I have a theory circulating out on the periphery. It was worked out pretty well I thought in another book *Knight of a 1000 eyes*. There, I was exploring the binary tree and the tesseract. I did a mapping of the primary processes of personality, as expressed in Jung and the Meyers Briggs Type Indicator, showing that it could be seen as a binary tree and also as the nodes of a tesseract.

Now I want to try and see if I can use the methodology of Levi-Strauss to see this same binary tree structure in a chain of analogies that indicated the ratiocination of the savage mind. That is to say, is the Totemic Operator a Tesseract?

It's not really a theory. What I have is a set of behaviors centered around a desire to see into higher dimensions. I like looking for models to give me some experience doing that. We will be doing an action — Gestalting the Tesseract: Line is to square as square is to cube as cube is to tesseract.

Line : square :: square : cube :: tesseract

The *bricoleur* uses whatever tools he has at hand to solve a problem.

Quasignosis : Ignorance :: Hypnosis : Sublimation

There are periods of weeks during which I go on a binge, trying to get some insight into the aesthetic problem I have

set myself: How does one think about and write about higher dimensional being.

Why would one want to do this? I tell myself the reason for all this imaginative work is because I think it might be possible to feel more free in my time if I am not being pushed around by all these desires. Or it might be because I have a fundamental need to transcend the nihilism — into a more positive, feeling perspective. It might be because I want to understand my mind. But way in the back of my mind it was because I needed to come up with some suitable ending to this book I was working on at the time, so I can get on with my life. (Launching myself into this book seemed like a good idea at the time, but now I was starting to feel beset upon, bedraggled, also tormented and tantalized as well as harassed and harangued. All this leads to the free running gestalt of quassignosis. I have a tiger by the tail and daren't let go lest a terrible mauling be perpetrated on my personal ego. Moreover, all this mental activity, though not as strenuous as mixing concrete with a hoe, is tiring and distracts from dealing with the life situation nonetheless. This indecision is reflected in my art.

Consequently there are periods of slipping into Quasignosis where I am unable to make the needed changes in the document because once you reach a certain level of order and symmetry, new perturbations to that order cause a lot more work, some of which proves to be unnecessary and rooted in a weak sense of self worth. Other reasons for the onset of Quasignosis are: because thinking is hard; because no one understands anyway; because it is hard to get back going once you have been thrown off the track. But ultimately the reason for this aesthetic journey is because writing and theory-building fill my life with intellectual rationale and because I get flashes of playful insight and epiphany that enlighten my path and give me a sense of fulfilling the purpose of my destiny.

However now because the outside indecision has manifest itself in my place of order, my art, all I can do is slip once again into one of those "deer in the headlights" dazes, Quasignosis, when you see the train wreck coming and just can't move to get out of the way. In order to distract from this feeling of helplessness I want to fill my useless waking hours with juvenile sex, sleep, get wrapped up in useless projects like taking care of the house or working for money, or get high and forget about it when I should be reading theory and trying to work it out. Such distractions help me forget that I'm supposed to be thinking about the exploration of higher dimensions. Eventually, however, anxiety about my worthless behavior forces another wave of desperate theorization.

I was more or less in this state when Laura had a huge fight with her mom, and got kicked out of the house, or moved out, and started staying on and off with me.

One evening Laura and I were playing ping pong at Austin Rec Center nearby on Rio Grande. She was talking about how a lot of her friends were getting accepted at college. She seemed agitated and troubled about what the future would hold for her.

"You know when I was your age, I wasn't going to go to college. I wanted to go up to New York city and see some beatniks and listen to jazz and become a writer. I had a job in Canada at the Greyrocks Inn outside Montreal. I got to ski every day. My only ambition in life was to be an accomplished skier. I was working as a waiter and living in the Staff Quarters, a nice modern house / cabin across the lake. When the lake froze over that winter I used to walk across it to work. It was beautiful but I wasn't meeting any like-minded people I could talk too.

"Anyway, my point is I think you are a very intelligent person and you might get bored in the work world. You won't meet people you can talk to. This hunger to be with

like-minded intelligent, idealistic young people will make you desirous to join into the college experience. Then you'll really be ready for college."

I was trying not to sound like all the controlling, pontificating, parental figures who were unable to talk about anything else except what her future should be like. She was getting enough of that. I told her right out, "I am not big on school. I wouldn't make a great guidance councilor."

"Anyway, when I came back to the States from Canada, I was eligible for the draft, so I got into San Antonio College to save my life. And that's how I came to go to college."

I hammed it up acting like I was Blind Justice holding the scale: "It was a choice. On the one hand I could let myself get drafted, and get my soulful, beatnik, ski-bum personage sent into hideous violent jungle warfare in Viet Nam, where I'd probably get my head blown off and maybe re-sewn on — and I would have to have a big scar around my neck, like Frankenstein for the rest of my life. Or on the other hand, I could get enrolled in a few classes, and play bridge all day in the Student Union building and wait it out. Plus, in those days UT was only $50 a semester. And you could get in easy."

She seemed to enjoy this presentation of an alternative perspective.

I continued a little bit more. "I have to chuckle at the memory. In my first English composition class, I chose to write a book report on Sartre's *Being and Nothingness*. Quel hubris!

"I remember getting really all worked up about Being and Identity. In existentialism there were all these different kinds of Being. I was going around every which way from sideways with all these terms. Being-for-one's-self. Being-for-others. I must have sounded like a complete idiot, spouting off these terms, in that crowded English class in San Antonio College.

And then, reading about all the nothingness in the existen-

tialist books. It was more a Buddhist nothingness — but I didn't understand it that way. I thought it was a sophisticated, stoic feeling about the ultimate truth and my teen angst was getting amplified with all the nothingness and nihilism I saw around me. That time in your life is a big transition and I wasn't handling it very well at all. It's a stage where you go from the idealism of everything is black and white to the realism of gray.

"They had a whole bunch of terms for Being. I remember one, Being-without-thing-ness. I guess it was some kind of primaeval potentiality from which all things emerged into existence. I don't know. Of course nobody understood it at San Antonio junior college, either. I guess I was really hungry to understand something.

"Ah yes that book was tough. God I'd like to see my book report now. It was truly atrocious.

"And philosophy shouldn't be like that. It should be like: How to Have Fun with your New Head.

"A person could have this whole menu of philosophical states he could explore."

I didn't get philosophical with her in our talk. I realized for myself though, that I was starting to at least have some language to think about these things. Later I got the intuition into having a revelation of an absolute perception.

The Tesseract is one of these Absolute Perceptions.

I didn't dismiss what her parents were telling her. I told her that even though it was not what she wanted to hear right now, they have her best interest at heart. "Your relationship with them is changing profoundly."

I typed up a solo theatre piece I had started in Berkeley — *The Subliminal Kid*. It was about an experience I had here in Austin, where I worked at a sleep lab. I was an assistant

bench tech. I worked under the supervision of an old friend from high school who had become an electrical engineer, on the op-amps and other electrical equipment. I felt like a phenomenologist using these op-amp circuits of a kind of, sensitized matter to pick up signals beyond ordinary perception. In *The Subliminal Kid* I used the idea of encephalopods attached to the head, picking up brain waves and transmitting them to a computer. There they triggered accessing imagery banks and other stimuli that would be sent back to the person. This was fairly sci-fi in the late 70s.

Seems like I should have been able to get another job like that, but there was something bohemian, even lycanthropic in my demeanor now. I had become an outsider. Sometimes it gave me pause. I felt embarrassed at my worldly lot. At least if I had studied auto mechanics instead of quantum mechanics, I'd have a car by now.

Mythopoiesis as Echo Location through Dialogic Field of Four Elements

Good-bye Party

We managed to go to court and get another month extension from the judge. A sense of becoming rootless had settled in over some of the participants in the final, last, going-away party that the people of Baylor Street, the Tribesmen of Neverland, threw for themselves. That morning a dark, heavy, "sure is tryin'-to rain" cloud hovered, threatening. And then just up and vanished for the revelers. But not the anxiety of homelessness that was beginning to settle in — way in the back of the minds of some of the hill folks there. And it manifest in roving eyes, wandering wiles and roaming that they tried to mask as gay abandon. Laura was finishing high school, and about to get her own place and a job. All the denizens of Baylor were scrounging for new digs, better jobs to afford them, new alliances. I of course, was leaving things 'till the last minute, a cast your fate to the wind attitude that is perhaps the curse of the INFP intuitive personality.

There was no denying the impending diaspora now. It was sad and awkward, how the tribesmen were saying a good-bye without really ever saying good-bye. During the day Bernie had stoked up a big hardwood fire in the middle of the Baylor Street central courtyard. He harvested coals for the big oil-drum barbecue pit.

I liked to throw parties more than I liked to attend them, because you have a lot more mobility as host. You can cruise the place and check everybody out a lot easier. Wild Bill and me and Laura and Isabelle went and got the great keg of Coors and ice and 150 cups during the peak thronging of Pease Park that was going on for some big pagan festival celebration. The Chronicle paper later said 2000 people

consumed 194 kegs of beer. The winning costume was a woman dressed resembling a penis, complete with men labeled sperm who attacked and kept bouncing off another plushly-pillowed, buxom young lady. It won the most original costume award.

When we got back to Baylor Street, Bill carried the keg up the hill to the central courtyard. It weighed over 100 pounds. He lowered the keg onto a bed of ice. I took up the long syringe-like pump, removed a little plastic dust cap and pushed and sunk it through the rubber grommet. Looking up at Isabelle, I smiled as a few drops of the precious white foam spewed out of the nozzle. Sitting there, looking for all the world like the angel 14-year old girl she was, Isabelle had a look that said she was going to get drunk tonight and who knew what kind of great transformations she would undergo. Her smile, though not shy and innocent, was curious in that it was encouraging.

"It's a good thing we know a weight lifter," said Billy.

"Yea," a chorus of bystanders marveled at Bill's great strength.

The first glass of beer was almost all foam, and out of this just-tapped, shimmering fizz, the party began to flow.

I had just gotten off work, and cleaned up, and now wanted something to eat. "Who's in charge of the chicken?"

Laura went into the kitchen and started hacking them in half with a big knife. Billy and Bernie got to work on the barbecue sauce. I was glad to have a little time to myself, to get some composure for I knew it was going to be a wild evening. Laura, 18 now, and blossomed into a mature young woman with big strong body, healthy tanned skin, and fabulous red hair was ready to party though I still held on to being protective of her. Taking a girl from her father's house, is a not to be taken lightly. Like all young people there is a yearning for love beyond what love she might have felt as a

child, or seen with her parents, or on TV and the movies. She has to go through the vicissitudes of transference, to learn how to get that love for herself. Coming to own your own sexuality in love is not easily accomplished and is fraught with the pains of unknowing. She is not so easy with her sexuality and lord knows, neither am I, but she was very interested in sex, and I was glad she keeps coming around. I drank a beer. The poet Jim Ryan came up, and I told him I had not been able to cop a lid for him. He went off in search for some, saying he would be back. I drank a cup of strong coffee, and kicked back for a while in my secret roof-deck garden on the other side of the house away from the festivities.

A while later, I took over with the chicken. Isabelle and Laura were to be maid servants, and passed the first batch around for the immediate tribesmen of Baylor Street. My fellow carpenter Greg was there. Wild Bill's dogs, Griz, Caylef and Sissy were ecstatic at the prospects of all the chicken in the world to eat.

It was a good party. Mattresses were laid out. A smoky fire drifting — wafting the smells of roasting fowl across the central courtyard. About 100 people showed up. So much was happening. All one's lovely friends arrived but you can not have any kind of deep conversation with them. Hans and wife Marie were there with their kids Kirsten and Nadia. They were friends of Laura and Isabelle as well.

"Hans is here!" yelled Isabelle and went off to see him. She then helped the band unload equipment down on the roof of the garage on Baylor Street. The band cranked up — R&B, Rockabilly. Johnny and Sara were dancing around wildly. Then Sara was singing. She is a real soulful woman.

The young girls were wild. Laura was taking Benadrill for some bite. Trooper and wife Mary showed up, with Damion and his wife. We all sat around the kitchen table smoking some dope they brought with them.

Another round of chicken was quickly started by Wild Bill.

Dubbie who came with Trooper — I immediately took a dislike to him — slipped Laura a drug called Mandrax, and the little punk took it. It sounds like something you would use to kill insects. This really made a mess of her. She and Isabelle went around drunk and singing. Soon Laura was drinking out of one bottle with another bottle of beer in her pocket.

After the drugs she was staggering around in the bushes, falling off of the garage roof into the dirt, making passes at all the boys, and hugging on strange girls, playing footsie with them at my kitchen table. She began hugging on a strange girl that she obviously thought was someone else. I meanwhile, was getting expertly picked up by a fine older woman, Kim, one of the eternal 25 year-olds.

Wild Bill started a third round of chickens late in the night. Kim was getting ready to leave, I had been chatting her up alternately with being the host and running heard and being mother-hen to this whole troop of 14 to 18 year old girls who were getting damned outrageous at the party. Even Hans's daughters were getting a high, and they are of very sedate German stock. As Kim was starting to leave, she and I were standing there surrounded in the drunken carnage of the 11th hour. We were alone together, adrift under a fish-shaped kite blowing in the breeze and I just pulled her into my arms and she demurred and let him kiss her and hug her. She smelled so nice, and looked good in this flowing Mexican peasant ensemble with vest. She is a well built female, with long dark hair down to the middle of her back.

I walked her up to her car; she did not say good-bye to Talbot because young Wendy was resting in there. Wendy had undergone an abortion operation early that day. Kim said it was her second, and Kim thought there might be some complications. I cringed with male guilt: these women were too

young to be put in that situation, damn it. So when we got up to Kim's car and I kissed her good night. Then she asked me to come along home with her to the farm. But I was thinking it was my responsibility to keep an eye on these young chicks in the hen house here and said, "No, there is too much I have to attend to here." I thought about seeing if she was interested to make it with me in the front seat of her car somewhere, but that seemed a little barbaric. Then I remembered my philosophy of never, never refusing a willing woman, and so went with her way out to her house in the country. I had been out there once before when Talbot introduced us, she was an old friend of his. Her place is a wild-looking German gingerbread farm house of the planes, a big roomy house of the Texas prairie, out near Round Rock. It is shadowy and mournful looking, on the moonful night — all set about with tall cypress trees in a lawn behind a gate.

I felt on top of the scene that night. It was cold, and we jumped into bed right quick. I got her all hot and bothered and she turned out to be one of the most responsive women I had been with in a long time. She was very sure of herself. She pulled off my pants saying, "Do these buttons really work?" She had a blasé humor about her sexual expertise like she was some kind of sexual sophisticate, which she was. I was proud of my 'cheap-ass pants' I got from the Goodwill with fake buttons down the front of the crotch; I liked their match flair.

We pleasured each other with oral sex. I was really turned on, and entered her. She pushed me off and said, not to because she didn't have a diaphragm. And I said, "I just wanted to play around." After a while, she came and she said, "I want you to come inside of me," and got up and put this big diaphragm in. The infernal thing looks like a piece of pink inner tube. Then she got on top of me and rode me like I was some kind of wild steed. I liked being ridden in the woman on top position a lot, because I can stay with it for a long time

while they have several orgasms. I can remain hard, just really digging on the deep pumping grind. In fact she was riding me madly close to sheer distraction. I began babbling in a devilish sneer, Fuck me, fuck me. I made sweet moan. Then after she came some more and was tiring out I started putting it to her really hard from below jerking her hips forcefully back and forth. Then I reared up, and putting one hand on the back of her head like I was lowering a new born babe into a cradle and was solicitous for her safety, I lowered her down under me on the bed and really started throwing it into her. Orgasms racked her in sequence coming rapidly one right after another separated by a few hot rushy breaths. She was delirious, writhing under the nuts, and calling me, "Honey."

Nothing like friendly pussy from the eternal 25 year old. It was true: Girls don't learn to laugh or fuck until they have suffered a little and are near 25 years old. Before that age they are eager but useless.

When I got back the next day, there were all kinds of teenagers standing around. Isabelle was still quite out of it, wandering around looking for her shoes, after having spent the night in the tree house with Mandy's husband Roberto. Wendy, apparently bounced back fine from her operation and with all the gusto that youth can muster, picked up Juliette's old man Peter! And Laura slept with some guy perhaps Dubbie on Talbot's back deck! At least that's what Isabelle told me. She said she saw them naked, at least partly naked on that back deck. But Laura had spent the rest of the night alone in my bed upstairs. So Isa might just be exercising her manipulative bitch powers too.

No doubt there would be some chameleon blushing later. There was enough rueful repercussion to go around for everyone.

Heavy Metal Music at the Steel Works

One afternoon Laura came over with a large gram of mescaline. We tripped together, and I made sure she felt comfortable and empowered. We saw lots of interesting things, especially the aleatory symphony sounds coming out of Tips Steel Works down at the end of Baylor Street where it meets the tracks running along the river.

Now from experience, I know mescaline is solar and is about forms. Whereas psilocybin is lunar and is about projections. Peyote likes to resonate with the great transcendental forms streaming down from the archetypal source represented in Old Sol. Mescalito is a champion of the reason and order behind the unfolding of things as they are. So if possible I like to be sure to trip on mescaline in the daytime. On the other hand, mushrooms and LSD are lunar and shadowy. They can make night out of day, and take the attributes out of forms. The unconscious comes to the fore in an aura of chiaroscuro, of personified entities cavorting in the half light of dappled gloaming underneath the boughs of a forest in the night.

Laura had written a poem to me that to my eye had quite a Shakespearean tone to it. I think they were studying that in school:

> Your unbashful forehead shines in the moon.
> Tell me of your antique world and teach me your low content.
> Live in a coat, and in the bounds of feed,
> Let the cows sprout mushrooms.
> I, despite my inventions, I dip cones continuously in my dream, a discord in my spheres.
> Did you hear the squirrel cats tearing flesh early this morning?
> The lustful sting of your love is swollen in your pants.

Pretty nice. I liked it. It was cool. Squirrel cats is the term

Billy came up with for the numerous feral cats that scurried over the arboreal highways of Neverland at Baylor Street. Also Laura had a job at an ice cream parlor on West Lynn, at the edge of Clarksville. Dipping cones. It was really sweet. I would meet her at the Clarksville Creamery to walk her home in the dark when she was closing up. They stayed open late as it got further into summer.

Another thing about psychedelics, is that it gets my mind so speeded up and convolved that I get too distracted to be able to give lovemaking the attention it deserves. Which is too bad because I have heard that women, on the other hand, quite like making love when their natural agape is amplified by psychedelics. But for myself, I get so far into mind, (after all you get access to the deeper ancient brain functions) that I become too transcendentally distracted to manage even a quick juvenile boff.

So we are off on a walkabout in the day. Two souls, who have gotten to a good place of trust with each other ready for the adventure of meandering the streets of dear old west Austin down to the river, keeping an eye on each other.

In the land of plenty, the sirens wail, tires squeal.

I feel like a big head floating loose of its tether to the body and floating out over the city, high in the sky. There is such incredible air traffic, these airplanes are swarming all over the city, swarming in my head one with a light flashing some kind of a slogan — the drone of it's motor like a koan, as it circles in the light cone of all it can reach with sound, plopped like an ice cream right in the middle of the sky.

Mescaline. It makes me feel the body in which I dwell, for it unfolds in me like a flower. It reminds me of that He that has been given me, for a while. It allows me to know

my soul and see the faces I put on, like masks in my run. It gestalts the guardian — that ego, or super ego *in loco parentis* watcher, so that I see how he becomes open to me and then closes to me in my confession. This guardian bent on flinging out all that is dissimilar to it, yet who somehow knows the correct name of things, separates and divides like mitosis meiosis or like Moses parting the sea, a slick trick, a ride that is not even of this earth alone. It is the elixir to lighten the vast emptiness of the void. It bewitches us with its light. This is what alchemists have sought; saints have sought it too. Poets have tried to w(rap) their words around it, mad men have seen it in the bursting bubble of the sphere of paranoia. There is no rest once it has been given you so I like to get walking, on a great long walkabout — in which there is no past but also from which there is no fatigue, nor will there ever be for there is no time. I wonder: Oh what is it that has been given me which has set my brain on fire, and my feet flying on pilgrimage, my nerves in a state of entrainment to the synchronistic. It makes this observer an interloper without guilt and no longer governed my time.

Down at the end of Baylor Street where it crosses the train tracks along the river we stopped in the shade by the Tips Iron Works.

There I could hear the foundry playing a heavy metal cacophony symphony in the banging and clanging, grating and chain slanging rattle.

I took my girl's arm and said, "Lets sit here in the shade of this old building and listen to the incredible clamor of sound coming out of the steel plant."

In the afternoon heat, the men beat the steel, and such a cacophony. The big girders going bong in the cavernous warehouse played the deep stuff, and the clatter of sheet

metal plates and the banging of steel tables play notes much higher above the grinding of the grind stone. The whole building was an orchestra in a great cave of the adventitious. It was the abode of some unholy djin within, and we slipped into the shade for a while and enjoyed it. My Laura is sweet to indulge the weird poetic ways of this old head. I could hear the chorus of Pythagorean harmonies in the cacophony. That old steel mill was a player of aleatory sound. Oh stochastic Stockhausen! In the Moment listening in to the *sturm und drang* of the press gang banging on the main line and other parts of the building around there getting their licks in too.

The great deep hum of engines combined with the drone of compressors and hydraulic pumps to sound like a chorus of monks OMing amid the ping, the pop, the plunk of all that heavy metal junk and the screech of the grinder to teach smooth speech.

Pound pound pound pound went the girder rounder.

I felt my mind getting analytical, trying to associate and ratiocinate out melodies from the heavy metal mash-up. I was having a Pythagorean moment. He must have felt like this hanging out besides the blacksmiths door listening to the workers inside the cavernous cave as they hammered big and small on anvils. It came to him how the notes were related to the energy of the blows and the size of the thumping hammers and the speed with which they fell. From there he got into bells and glockenspiels and strings of lute.

"Wow listen to that," I said.

It was some kind of tremendous heavy-metal malediction. The growl and jangle in the mangled tangle of steady whirring punch / press machines was playing with the turn / pull machines. The pound / cut machines were jamming with the angle cutter and the helical bore. All to the beat of the great big banging bong, pounding out the measure of the age-long

love song of the steel men. Then the high sibilant chatter and breaking splitting splatter, like the whole building was about to explode, battered to tatters as the pieces fell and clattered into the crux of the matter.

Meanwhile in the hollows of the cavernous building, reverbarians were fighting barbarians while in from the skies came riding mongol hoards of Valkyries becoming the bloom of the whom swirling around the chaos and the mighty din of change and deconstruction and reconstruction.

"What an amazing Heavy Metal concert!" I exclaimed, for I wanted to continued to tarry in Pythagorean bliss, a little while longer. We hung there quiet together amid the great riot — this holy goof in the warp and the woof, of the aleatory harmonies — like two little mice without the admission price, enjoying the concert through a hole in the wall.

Later we walked along the river, holding hands.

Back at my soon to be vacated flat, I looked out from the deck. I saw six werewolves sitting in my biggest pot plant.

"Damn it," I realized — "I think it is going to turn out to be a male, got little horns growing on it already."

I am able to see myself behind the frieze of pot plants, curly hair, glasses, enthusiastic, intent. I look out at the city and wonder at myself, and think, what the hell. . .

I was able to get Kim to let me transplant the remaining females out at her place. Her roommates wanted to learn the art of raising and distilling spirit.

Epilogue

Wild Bill moved back into his old house on Theresa St. in Clarksville with the roommate his wife got to replace him. She went to Oregon to become a nurse. I moved into a place also in Clarksville, down the road (unpaved) from where Wild Bill was staying. A woman whose husband had been sentenced to jail needed someone to help with the rent. She had a kid and a dog. After a very short while she had to go live with a relative and I inherited the dog, Sunshine a sweet pit bull. It was good for me to have this being to take care of, and since I had already started taking Sunshine on the long walks with Wild Bill and his three dogs, it worked out well.

Laura got her own apartment with a girlfriend and we continued to see each other throughout the summer. I'd walk her home from her job at the ice cream parlor in Clarksville. I got a suit and snuck into her graduation at the coliseum on the river. I did not go introduce myself to her then, because she was with her parents and her classmates and I thought it might be awkward. It was my little secret homage, and I just wanted my good vibes to be in the room. I got her a little necklace with a cross. She was very sweet to me on my 31st birthday.

I had to get out of the rent house I had to take over from the jailbird's wife — money issues and they were paving the streets in the heat and it was hideous. I moved in with Wild Bill on Theresa St. I stayed there quite a while on a couch in the living room. We had a lot of fun. We would try to be in the house at night by 11:00 for a very cool DJ to open his show with the tune *Pauli Gap* of Jimi Hendrix. This soulful blues / jazz tune was a masterpiece that sent my mind drifting into pure rhapsody every time I heard it. By then he knew everything there was to know about soul.

Appendix:

A Structural Analysis

— A selection of essay-like pieces being worked on at the time. Included here because they extend the novel's communication field elements from words to matrices, tables, diagrams, scattering trajectories, circuits and images, and would have disrupted the normal narrative flow of the story.

Introduction	217
The Sorcerer's Apprentice	219
A Toast to Carlos Castaneda	235
The Theatre of the Tesseract (Intro)	238
The Subliminal Kid (a monologue)	241
The Theatre of the Tesseract (Notes)	262
Quarking the Cube	285
And we are way downstream in the light	313

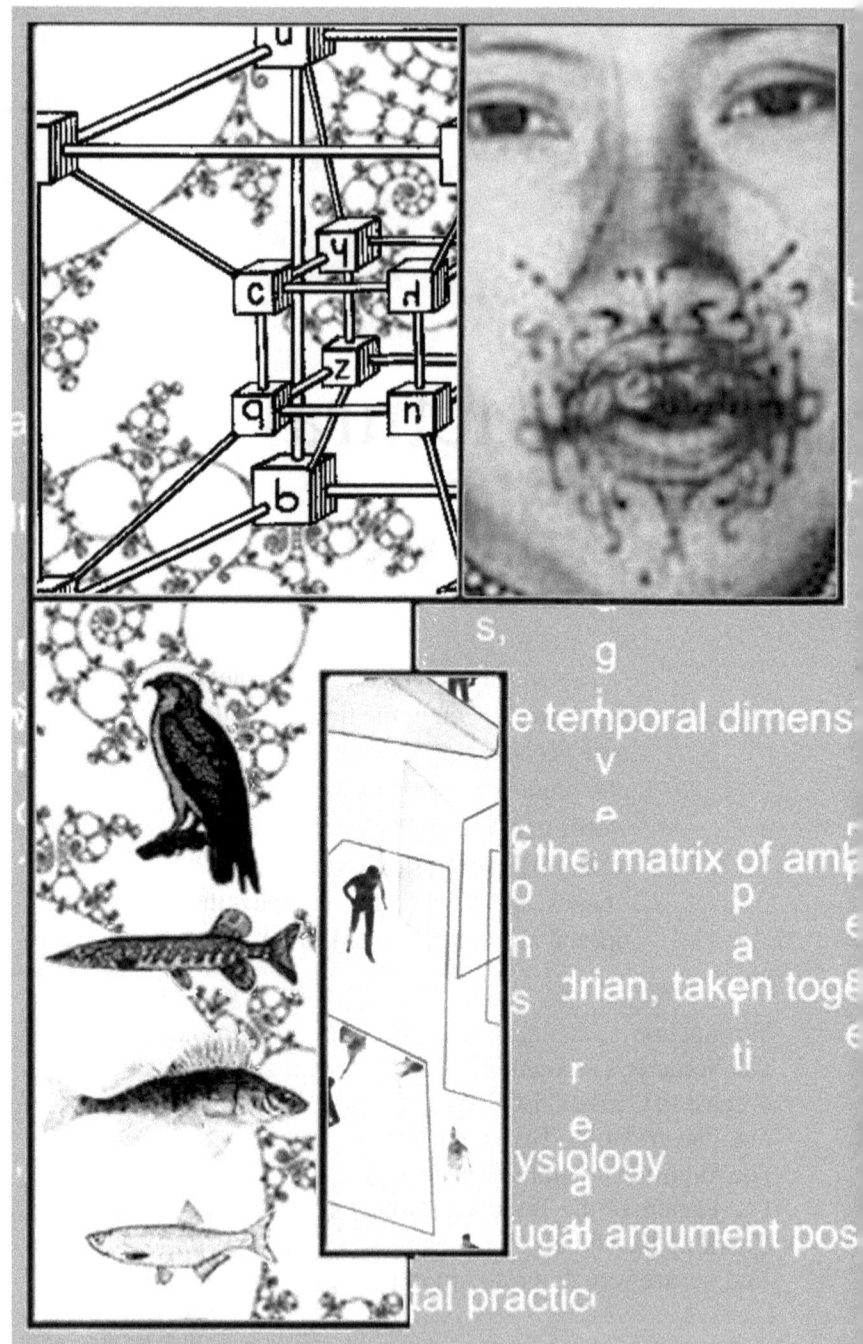
e temporal dimens
the matrix of amb
drian, taken tog
ysiology
ugal argument pos
tal practic

Introduction

This book explores meanings that "the Indigenous" and "Neverland" have for me. There are at least four associations: 1) the structural anthropology of Levi-Strauss, in particular the Totemic Operator; 2) the archetypal psychology of Jung, in particular the Peter Pan Syndrome; 3) the physics of Einstein and Heisenburg, in particular quantum Bose fields; and 4) the Buddha lands.

Neverland is associated with Peter Pan. It has come to symbolize that time of childhood when the mind could be deeply engaged in fantasy, working out what was going on from what little information is available. Many papers have been written about the Jungian Archetype of the Eternal Child. The Peter Pan Syndrome came to be a derogatory term for certain young men — romantics, bohemians, hippies, who refused to grow up and become enslaved in consumer society.

The Indigenous is associated with tribal, natural, culture. The anthropology of modern consciousness starts in the study of myth. Structuralism has much to teach us. Semiotics.

I also associate Neverland with field quanta. These mediate how we sense the world. These are massless particles: photon, graviton, neutrino etc. carrying information and flow through anything. They 'never land', and never move slower or faster than the speed of light. This surface of the Light Cone, a hypersphere, contains phenomenal reality.

Another association of Neverland is the Buddha Realm. That psychological religion describes various pure and pristine "lands" that they are trying to get to in their being, with their practice. I think this refers to a state of being able to hold and conduct light. But it is basically those rare moments where you are able to wake up from all the thoughts and worries pushing you around and just be in the generosity,

(which really is splendid.) That sense of well being, and knowing it, is really what humans are yearning for. It is that place, I called Neverland. Once you've had a taste of really feeling yourself blessed in the generosity, you will want to be in that feeling more often. It could be the common everyday trance. You know — that reverie you slip into when stuck in traffic on a rainy day. Or sometimes you just look up at the clouds moving fast through the sky on a spring day when the world seems so fresh. You just want to let your attention come to rest in awareness. You gotta have chaos to get pulled in by the attractors.

Through myth, Jung and Levi-Strauss and Campbell and Joyce, have shown us how consciousness is changing. Mysticism, once the prerequisite of only a few quixotic, spirited, sublime souls, has in one generation become a necessity for many. Yes, we are all trying to get into Neverland in our way.

The ways into Neverland are many: archetypal psychology, personal mythology, anthropology. The study of Neverland gets into quantum field theory, matrix logic and group theory for the field part but I won't get technical. This is still a novel. Let us explore then, into the aesthetic psychology, the epistemology and logic of our time.

Many papers and books have been written about the mythology and psychology of the indigenous peoples of the earth. Levi-Strauss and others were trying to capture, before it becomes extinct, this ancient logic of the concrete and the body which is more pristine and pronounced and alive in the myths and culture of the tribes. The totemic can be encapsulated in the dictum: Animals are good to think with. It speaks to the eternal rift man feels from his animal nature. Though myth emerged with homo sapiens in the paleolithic it is still very much a part of the modern mind.

This appendix is a kind of auto-ethnography, more or less in the order it presented itself.

The Sorcerer's Apprentice

It has been eleven years since Carlos Castaneda first opened a portal to a world parallel to our own with his books starting with *The Teachings of Don Juan: A Yaqui Way of Knowledge*. In these tales of power and self-knowledge, Castaneda changed the face of anthropology and moved the human potential movement to the front of popular culture. He gave the hippies an avatar of their zeitgeist: don Juan Matus. Now, there is debate over whether it was real anthropology or analogical fiction. Though the makers of opinion now have withdrawn their support for the work of Castaneda, the media at the time made it a big part of the popular culture then. Like everyone else, they believed or at least went along with it, because it was so *zeitgeist*. At least among the tribe I hung out with, it was of interest. But at the end, the further installments of the franchise are getting too far away from believability. Castaneda did a lot for making anthropology and ethnology and the savage mind interesting to the average public. His writing, in a story-telling style, was also committed to philosophy and psychology. The intellectual rigors of Structuralism and Semiotics, and the sociological writing of ethnology, were intimidating; but in the writings of Castaneda, the western student of mystical experience found a kindred soul. Especially if you were a writer: one had to admire this guy going around with his notebook taking down what was going on.

Like most young seekers in my generation I enjoyed the amplified teaching of don Juan and the reactions of his intrepid apprentice. It came at a time when everybody was trying to find their way to an alternative reality, something different from the terrible conscripted war in the jungle, and from the cold war; something more natural and real then the suburbs. This writing

fueled consciousness expansion. We were hip to anthropology, to all these explorers going into primitive uncorrupted cultures through their art and through their ceremonies and kinship structures, trying to find our way into the innocent childhood unconscious of man, trying to understand ourselves by going back to the source. As a student of Levi-Strauss I found it quite a struggle to get through structuralism, and semiotics. These academics and critics of culture could be annoying with their sour, arrogant, French rectitude. They studied religion and philosophy in grade school and it made them a very serious bunch. In Castaneda, it seemed like one could just take a drug, peyote or mescaline or psilocybin and get right directly into the experience. If one could take down the filters of socially constructed reality one could get an intuition into the universe.

Reading Castaneda, you were brought into another world. It felt wondrous to be more attuned to this ancient world, this non-ordinary reality you had entered with him. Suddenly your senses were sharper, you felt more alive. I recall feeling something like chakras or evolutionary awarenesses. I tried to visualize my aura as a luminous egg surrounding my physical body, I walked taller trying to get my will aligned with the Point of Attention floating somewhere in the air straight up from my spine. It was fun to walk around trying to sense these metaphysical organs and appendages and imagine new ones of a future people made possible by the discoveries of today. It would be a great advantage to know them. Science is the new religion, and like the old religion it often served to keep truly spiritual / scientific people from practicing their beliefs. Here *beliefs* could be read as *hypotheses* for the scientific.

It was very inspiring to feel like you were the chosen one to study all this Man of Knowledge information. I realized, like Castaneda that my Ally was my writing. I felt like I was one of them. This is my confession.

"Oh. You've been a an untrustworthy narrator."

Forgive me, my brothers and sisters reading this novel, for I have sinned. It has been 7 years since I started reading Castaneda and I must confess that I, in the early days, thought it real. Or at least probable. And if not, then at least it was a committed fiction.

In the summer of 1973, I was just out of university with a degree in science and should have been much more of a pragnatuc realist. (We'll have more to say about that later.) I was a Castaneda fan. I bought the hard copy of *Journey to Ixtlan* at the UT Book Store. A forklift had dropped a great wooden pallet of them, a gross of the books, upstairs. These hard bound books with a colorful jackets drawn in the style of a mystical comic. People were buying them up. I dropped everything and read it straight through. A couple of days later, I walked the earth looking for Death over my right shoulder. I wanted to get in a car and wander the great Sonoran dessert looking for a shaman to study under.

I wondered, was the writing allegorical? Was it real? I had read about Maria Sabinas in the gloriously illustrated editions of Gordon Wasson. I tried psilocybin and LSD with my girlfriend. I was about integrating this supposed ethnology with other books: *The Tibetan Book of the Dead* of Timothy Leary. And with Henri Mischaux's *Miserable Miracle*. And through Jung and Campbell my knowledge became acquainted of the occult universe. This feeling of being a part of some great Human Potential experiment was in the air. People were trying to break on through to the other side.

The world seemed filled with unlimited possibilities. (I'm talking novels, too). But then, I had come from physics, which has the keenest conceptual constructs, analogies and metaphors for these spaces wild and tame — that man has ever devised to foster savoring the generosity of the movement of energy through the forms of the universe.

Phase space and Hamiltonians of energy and Heizenburg matrix algebras, time dilation, time series, entropy, stochastic processes. I was constantly amazed by the exotic creative idea content in Physics. We made holograms, split a laser light beam on an interferometer, watched a spherical oil drop through a telescope glistening like another world as it negotiated the fields of gravity and electricity in a viscous flow. I engaged everyone who was game, in conversations. One day, I could understand what was going on in every star of the universe: a thermonuclear explosion fed by and contained in gravitational collapse. I had listened to and seen the background white noise left over in space from the Big Bang; or worked with autopoetic chemical reactions forming from lightning into the protein molecules of life. Ecology and life was ouroborosive like that too. How wonderful that the forms of the animals and plants were elegantly adapted to energy collection and distribution and recycling.

There were times when I felt myself to be just another Random Walker in this, The Age of the Stochastic Man. Or was just one of the many automaton machines communicating. Certainly we are all possessed by Maxwell's Demon. Now THERE was a myth of the new age: the sorting agent at the crux of information (anti-entropy) and energy.

Physics was my William Blake. It taught me to see a world in a grain of sand — the crystalline structure of reality, especially group theory and symmetry. It taught me to see heaven in a wild flower — the elegant surfaces of differential manifolds, the foliations of convex hulls.

I have held infinity in the palm of my hand, by zooming out in powers of 10, to looking at myself holding out my hand from the sky. Then from a satellite's view. Then out and out — past the planets to the galaxy and the clusters. Then zooming back down into the veins of my hand, into the blood in the veins, into the molecules in the blood, into the atoms in

the molecules, into the elementary particles of the atoms.

And I have beheld eternity, in an hour. It is what drove me to become obsessed with the study of time. I came to understand that the hours we watch closely are being watched through us by an eternity. Eternity looks like a vast sphere or hypersurface of many dimensions. There are different kinds of time, the parameter-time of the equations of motion of physical reality is only one type of time. There are many subjective times which are more like fractals, iterating forms, self-referentiating frames.

Physics was my Blake. Especially when you start to enter into the non-linear. The linear calculus of functions got into everything as long as it could be approached as linear. But now Poincare was leading the way, into the dynamics of the non-linear, as were Hilbert and Einstein. And to study another construction of time, was the beginning of the project for our generation, to take on a deeper responsibility for their personal consciousness. It was the zeitgeist of the times to study the recursive analogy mediation of a time-binder becoming a time-player.

The Indigenous Tribesmen of Neverland were trying to understand or feel the underlying connectivity that through all is interfused. It is the thought of the world mind, which moves on objective and subjective levels. It is thought trying to understand or feel the way a holographic matrix of light is interfused through all. Like the way light is to shadow; or being is to non being; or ocean is to sky. And the mind moving, as current, or wind, or myth, or meme, or music — connecting all things thought — up into a universal mind.

A mythological synesthesia, a music of meaning.

In quantum mechanics I got an idea how everything was based on probability, how the laws of the universe were symmetries of space and time and rotation and other groups.

We began to *see* in the idea of classical analog, like

poets. We began to see, how some thing, some form, some system was isomorphic to some other thing, entity, flow, or pattern. And then we see through the anthropic principle. It is the universe at work everywhere, a universe that has made all the right moves, to get us here. Here in this planet, in this galaxy, we came to understand how miraculously unique and blessed with fortuitous elan we were.

I attended the lectures Prigogine gave at UT. I liked him because he was into time, had read Proust and Borges and his idea that entropy required local pockets of complexity (life) in order to maximize entropy production seemed like the answer to the most fundamental question.

I saw how to model the ecology with electrical circuits, and understood how we were energy in circuit. There were so many amazing analogies and great graphical gestalts to be gained from physics. And the experiments! The dust motes of thermal agitation of atoms moved by the mythological agency of Maxwell's Demon. Now there is a story you could sail out on forever.

I would be walking around saying I was a citizen of the universe (an idea I first encountered in Joyce and Henry Miller.) I could be going over the thoughts of Von Neuman and Prigogine and Weyl and Dirac and Feynman, and the old guys Grassman and Hamilton and Poisson and Lagrange and young Galois, these Pythagorean saints and many others, G. Spencer Brown, and Carnot and Fourier with his ability to transpose everything into vibrations and Maxwell and Einstein and Minkowski and Klein. They had imparted their essence into my brain, had infected my brain with their great moving insights in their elegant theorems and intricately worked out expressions.

I was awakened by peyote, but it was the idea content of physics that perpetuated the awakening, even more than literature. But in my blind admiration of Castaneda's oeuvre

I was guilty of overlooking that even though his teacher had forbade him from using a tape recorder, the dialog was perfect as well as Socratic. Though there seemed to be quite a few special effects in terms of what they were seeing, the storytelling and pacing were certainly way more than that of a beginning recent anthropology graduate.

I wanted to astound myself by trying many of the student apprentice sorcerer experiments, like the stereophonic whispering in each ear to lull the mind into a phase space of semantic differential. Or to turn around real fast and try to see my Death over the left shoulder moving in some parallel universe. How I envied my more spiritually adept and phenomenologically gifted brothers and sisters who could see auras and know things about and beyond our ordinary reality. For me it was necessary to use science. I had to develop some kind of holographic model, imagining a storage medium in which every part carried the condensed-in-light information necessary to reconstruct the whole image from the coherent in-phase laser light, which is to say out of the harmonics of energies. (This idea of thinking about heat as spectrum rays was in Fourier, 1827.)

But my brothers and sisters, I am most ashamed of how I believed all of it — until when Don Juan appeared wearing a suite. That really over did it for me. Carlos was having us on. And always the melodrama about preparing for the onslaught of occult experiences.

I could just say, I was taken in by an artist. But the academic establishment, when they caught on — Oh how they took umbrage! At being a part of veracity deceived.

For many and for myself for a while there in the early 70s, the writings of Castaneda were the shamanic bible. The experiments that you could do on yourself opened the possibilities for the exploration of other realities besides the consensus one in which we find ourselves. Peyote was the sacrament in this

new anthropological religion. The writings of Castaneda filled a huge void, and helped us — through its detailed exposition of ceremony and behavior and structural analysis.

Both Levi-Strauss and Castaneda were talking about a mythological reality. Both were mythopoeic writers. Myth, that external structure that communicated with our unconscious, whatever IT was, was what we were experiencing on the trip. Mythopoiesis is at the heart of making sense.

Castaneda, after being Indianized by his guide shaman, focused the ethnological study inward on himself. He did experiments, conducted precise observations and recording, and pulled it all together in a structural analysis that was good enough to fool a university anthropology department into giving him a PhD. You've got to admire the young don, he got a PhD in his own stuff.

I read it as fiction. It used one of the finest literary structures ever invented: the old teaching the young. It was an allegorical tale on the surface, about the struggle of the old shaman to overthrow the tyranny of the western ego, as the story played out in the often ambiguous and downright antimonious relationship of the characters — the student Castaneda and the venerable don Juan. What is being played out here among other things, is an expiation of guilt. The story enacts an intricate and multilayered inversion, of the stereotypes given to us by our history. Now Carlos becomes the primitive in an alien culture, persecuted under constant threat of death. Indeed they used death as character in their passion play, standing over the left shoulder, somewhere out there in your destiny to propel you into greater existential experience.

Carlos undergoes a metamorphosis to the contrary: from the conquering imperialist representative of the white establishment, scorning the beliefs of the savage mind — into a disciple of this most perennial philosophy. It was a beautiful inversion of the racial stereotypes of history, one that we all

needed to see worked out.

Now I can appreciate it for what it was: the artist as ethnographer. Even though the preparations of each voyage into the separate reality got more and more outlandish and unbelievable, even though the structuralist analogy required study and thinking, it was an allegory of the psychic times we were all seeking in our lives, and it suggested ways in which we could approach this for ourselves.

Now, upon looking at my enthusiasm for this work, I begin to see the error of my ways. I was seduced by the temptations and distractions of the possibility of the return of the Paramedian Sphere — the *Naugual* to which all things aspired, whence they came and to which they evolved.

I confess that I did explore getting free of my personal history, even to the extent of leaving family and staying away. This was my great sin. I have strove to make up for it by *seeing* through literature and poetry and some kind of cross between poetry and algebra. And I have made much penance over this past decade through acts of Proustian heroism — going over and confessing my past life, subjecting my love relationships and self to a level of scrutiny they probably doesn't deserve.

My reconciliation would not be complete without confessing the shame of these last ten years. My Brothers and Sisters in the Church of the Coincidental Metaphor: I ask for your forgiveness. But I also ask you to join me in this cleansing rite. Step into the confessional. Open your heart. Admit that you too, read the tales of power and believed in the exploits of don Juan and his loving apprentice. And saw no fault in the books themselves nor the abomination of academic inauthenticity known as plagiarism. (At least for a few hours/days/weeks before you came to your senses.) This way salvation and forgiveness lie.

As much as I loved the apprenticeship of Castaneda and don Juan, it was not a path with honor, ultimately. It was hard

to admit, I had been taken in. On the other hand perhaps it is the usual shortening of the life expectancy of any reading.

What Castaneda did with this first don Juan book was what the philosophers of old had been trying to get the young to do since the crack of dawn — understand that the map is not the territory. One has to use these linguistical, artistic, spiritual, mystical, totemic structures to go beyond talk and beyond thinking and glimpse the living world on its own terms at the level of body knowing. Castaneda learned his anthropology by cleansing the doors of perception, and dissolving away the filters of preconceived ideas, all be it under the influence of a psychedelic drug. We recognized in Castaneda and Levi-Strauss, the work of the philosopher — taking apart our received ideas and unconscious assumptions about the world, that are contained, hidden away, in those maps. Castaneda — whatever he did — he for a while helped a number of people live a communal dream. One that forced them to, if not confront, at least be more aware of, consensual reality. These books were guides for approaching the world again with the beginner's mind of a child.

One of the big questions for Castaneda is finding a "path with heart." He asks don Juan how to know when a path has heart, and don Juan tells him simply, "Before you embark on it you ask the question: Does this path have heart? If the answer is no, you will know it, and then you must choose another path." That answer activates Carlos's fears that he might lie to himself. "Why would you lie?" don Juan asks him.

"Perhaps because at the moment the path is pleasant and enjoyable."

"That is nonsense. A path without a heart is never enjoyable. You have to work hard even to take it. On the other hand, a path with heart is easy; it does not make you work at liking it."

Castaneda described seeing people as the luminous

beings they are. It was like something out of a sci-fi thriller, with luminous fibers coming out of their middle. He described the Center of Attention as being up and to the back off the body. It sounded to me like the center of mass or centroid of a strange shaped body. This was such great stuff. I remember walking down the streets of Austin, with my spine more fully erected, and just *being* — in a very attuned place.

I made promises to myself at that time: I was going to find a path with heart. It would be like some kind of network or channel that came out of my chest and reached out to people of the world. In Tai Chi this is the center. You can focus energy on this place and project it. Also there were the Acting Exercises I did in Berkeley. The fact that I had spent some time with my head stuck in tomes of differential geometry and topology, helped with the construction of metaphor.

It was encouraging that the Man of Knowledge was chosen by an agent or entity of the spirit world; it wasn't something you applied for or wrote up a grant for. There are no overt requirements for becoming a Man of Knowledge; there were some covert requirements. The decision as to who could learn to become a Man of Knowledge was made by an Impersonal Power. The Man of Knowledge is chosen by his Ally to followed the path with heart as it leads on life's journey. No one knows how or why an Apprentice will be chosen. The Man of Knowledge is often blessed with a power to intuit objects moving in and out from the fourth dimension.

I felt like I had been chosen to be on this path. There was a certain feeling I had when I was in communion with myself that felt right. I felt lucky. I had no right to expect that somebody up there was keeping an eye out for me. That I was on a mission to become an artist since I was in grade 3. The artist is an antenna of his race. He is chosen to receive transmissions from the noumenal sphere and translate them through the interstices of the totemic operator into the language of the tribe.

The sorcerer's apprentice is aware of this noumenal sphere, the surface of which is ordinary reality. Or Newtonian / Euclidean / Aristotelian reality — i.e. a logical construct in which the phenomenon is observed but not queried as to its source. In physics this surface is that of the light cone of field quanta percolating off the real world beneath. I was blessed / cursed with a heightened awareness of time.

This got focused onto the idea of the tesseract, a 4 dimensional object. I thought that it was similar to Levi-Strauss's Totemic Operator, and to Jung's Archetypal Regulator. Other manifestations was the Sephierot at the heart of the Mystic Kabbala, and the Pa Qua at the heart of the I Ching. Another modern manifestation was the Tesseract in the MBTI. And another was the Semantic Differential of Count Korzybski of Non-Aristotelian logic. Other representations of this structure were the system of Correspondences in Baudelair and the Objective Correlatives in T.S. Eliot.

(The last paragraph would make a good PhD thesis; one could have a career in graduate school with it. But I didn't want to become an expert in this, I just wanted a useful aesthetic to make creativity easier.)

All groups have their totemic operator. We could even talk about it like an operator on a vector space. Levi-Strauss intuitively took this analogy a long way into penetrating myths. The totemic operator produces other myths, like a vector product, produces other vectors.

In the system of Castaneda, the universe is divided into two parts, an Ordinary Reality and a Separate Reality. Castaneda called them the Tonal and the Naugual. The partition between the Ordinary Reality and the Separate Reality beneath it (into which the unconscious has set its deep roots firing ganglia on the quantum level) is a place possessed of mystery and potential. There are guardians to keep the two worlds separate. These guardians have representation in the

human physiology as filters and in the world as archetypes.

These two worlds are manifest in all domains: the universe of matter and the universe of energy; or being and existence; or physics and mathematics. These junctions where parallel worlds become tangential afford a leap across. This is the generalized light cone, the situation of the event of now. Since the light cone is the reach of light, and therefore causality, the inside of the cone is the real, the edge or boundary of the cone is that neverland of light and causality implicating. The outside of this boundary is the beyond, the sum of all possible worlds. These coincidental junctions worked themselves out in the dialectic of binary opposites in Levi-Strauss.

It is the adventure of the Man of Knowledge to explore transcending the Tonal to experience the Naugual. The Ordinary Reality is a construct brought about by the severe behavior modification and brainwashing called culture, or upbringing or inculcation or childhood. Castaneda had various names for it, Consensus Reality is one. The world of Ordinary Reality is constantly filled with opportunities to see into the Naugual. These are called enlightenment experiences or insights or epiphany. Throughout his books Castaneda gave some principal techniques to stop or dissolve or transcend the ordinary realty: these were "Dreaming," "seeing," "stalking," "re- capitulation," "controlled folly," and "stopping the world." They are found in one form or another in most spiritual traditions and indigenous cultures around the world and designed to drop below the mask of our personality structure and access deeper ways of knowing the world and ourselves.

At one time or another I seriously practiced all these techniques. I did some of these aesthetic induced explorations with a friend, the painter Ben White. I remember doing the "gate of power" at night through the woods. It really worked, it was amazing to run at break neck speed over boulders, bringing your knees all the way up to your chin. It put you

into such a state of awareness that you did not make missteps and could change course in a split second because you were not committed until the last. One fairly flew over the rocks. Another exercise was looking for my hands in dreams, trying to learn the art of lucid dreaming.

Incidents of epiphany when the veil of Maya parts and lets your recognize the archetypal nature of the underlying universe are fairly rare for most people. They usually happened during life threatening occasions, but can also be induced by various spiritual, artistic, aesthetic practices. In essence you are kept from the unconscious by the set of filters that is consciousness. This is to protect you from being overwhelmed by the factorial dispersion of possibilities everywhere. However it is the responsibility of the owner to have fun with their head.

Best results in the breaching of the veil can be had with binges of creativity. A work of art is a vessel for piercing the veil. It does this by being the containment, the symmetry of detente for the opposing yet compelling forces on each side of the partitioned potential. These forces are coded in both fact and fiction. Visually they are a projection in which a dimension is lost. 3-D down to 2-D. Or in the case of music projection into a 3-D acoustical space capable of rhythmic entrainment.

Artworks are time portals projected into media. It was my desire to create texts that were time portals that can be used as gateways across times. Texts by some authors become tools that improve the way time portal construction is done. The texts become organizational gateways between universes at the partition between synergy and destiny, and are experienced as epiphanic. It is at these Sorting Points where a 2nd and higher order information demon is parsing. Castaneda thought of this too, the attention points.

The most import element in the sorcerer system presented

by Castaneda is the art of Lucid Dreaming. I pictured the dreaming brain mind as a tesseract of personality types.

Castaneda got us interested in Anthropology. When we came to anthro we found Levi-Strauss, and the linguists, Saussure and Barthes and semiotics. From Semiotics we found the American C.S. Peirce. That was one dimension. Another dimension was math. Groups, games, automata, von Neuman. And dimensionality itself.

Dimensionality is the connection of opposites along an axis. One dimension was my immersion in physics. We were in a time when people were discovering the underlying laws of the universe. And this was reflecting back into the structuralism of Levi-Strauss. Another dimension was psychology: the axis ran from Freud to Jung and somewhere in there was the Games people Play. This had rays leading into von Neuman, and Game theory, into Levi-Strauss the categorical system for being in a group or institution, Group theory for being in a group, and also we had this whole external operator system, outside the individual minds, Jung called them archetypes. Levi-Strauss called them Totemic Operator.

So there was a lot going on. I was lying in bed in that hypnagogic state between dream and wake and was thinking about an isomorphism or at least a homeomorphism between the Totemic Operator and the Tesseract. I wanted to take the MBTI modeled tesseract into one that expressed my state of being pulled in many directions, here in Austin.

To be in the Neverland of Austin was to oscillate, to gestalt, to shimmer between the poles of these dimensions.

From the Vector Space Theory of Matter of Matsen, to the Mahiyana Buddhism of Raja Rao, to the Kruppa's Electronic Literacy class there was a lot going on if you looked for it. Visiting writers, Borges, Burroughs, William Gass, visiting Mathematicians, Dirac, Prigogene. George Steiner, lectures on Freud, Marx, Levi-Strauss.

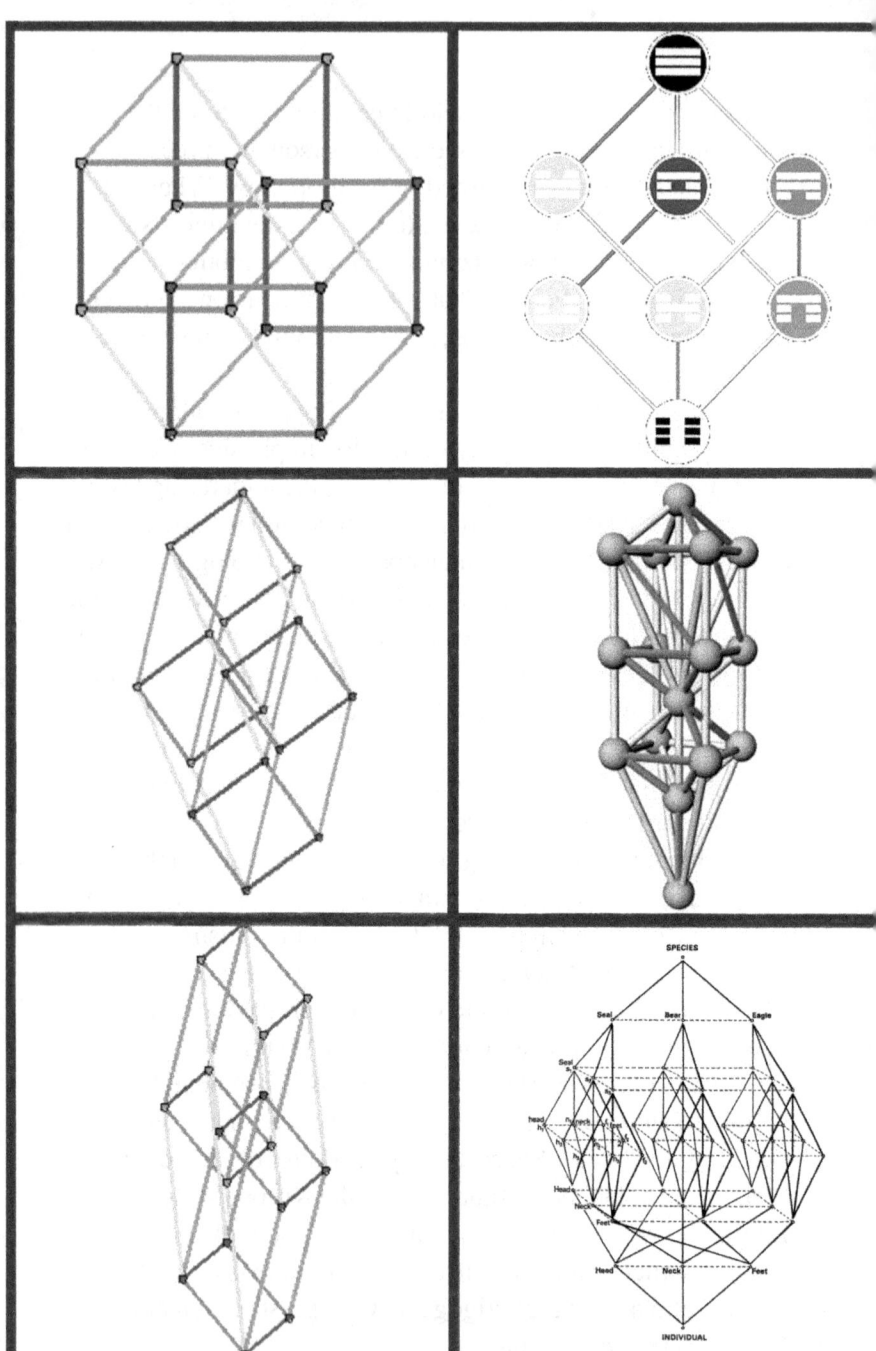

The Tesseract as Generalized Model for Archetypal Mnemonic Mandala

A Toast to Carlos Castaneda

Don Juan? Carlos? What a wild way that was.
I was an acolyte / attending classes when
the twilight sky
became an opening between worlds.
There we confronted our beliefs —
as we were confronted.
By a big black dog with phosphorescent eyes,
chasing us
through a mythical desert landscape,
its match-stick claws
igniting sparks,
in a stormy sky.
Or as an enormous moth, spread its eye-wings
across the setting sun,
we were deified
by the agency of knowing ourselves,
through the medium of this Other.

As it was,
the Mescalito, the Humito, the lophophora,
— the witches and other totemic operators
through which we pretend —the *bricoleur*,
we were initiated into the higher dimensions
and spent our youth, treading the line between worlds.

What a hunger for experience there was. . .
after all that time alone, with others in school.

But I always listened, through the border blaster radio,
— the sound, of screaming Jay Hawkins
and Bobby Blue Bland
and Wolfman Jack exhorting us in the night.
The sound washing over us in waves,
of energy / feeling / propagating at the speed of light
. . . somehow in a dream.

There were Time Animals moving through
like clouds, like the wind through fields
you couldn't see it
unless it moved against a tree or a flag.
Like shadows
— projections
            — scudding across the planes . . .
. . . leaving trails . . .

I have long hair,
so that you might feel yourself
confronted to look beyond the <Not Us> mode.
And it kept me out
of your market place:
I learned to live on found objects in another economy,
of Negative Capability
— halfway between Percept and Sign.

They thought that because I revered
revelations that I would allow
the special effects department to burry me
with car crashes between the commercials.

I have this habit, like Levi-Strauss or Feynman,
of looking for the general in the particular
and the particular in general,
and tightening up the leaky O-ring,
so that no one will get lost.

What was there for us, in that.
An encouraging ally? A tunnel between worlds?
A luminous being?
If I have lost the material desire
to be part of the status quo
then I have the responsibility to find
my own group and make it go.

I have been indigenized,
that is why I say this.

You Nagual, have sent out invitations to
the separate reality.
That analogy is above as below,
and is still the percept rising to sign
in our poems.
We savor it. And the people who have heard It,
take it home. And after smoking whatever it is, that
relieves them,
create radical fields
to extend it.

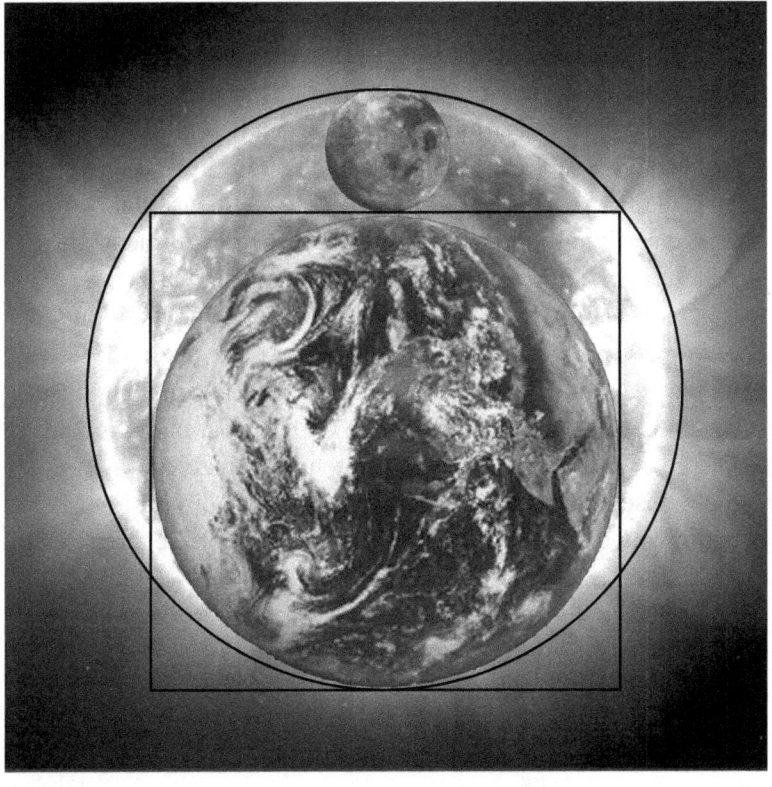

The Theatre of the Tesseract (Intro)

A Note on the Origin of the Theatre as Tesseract

I came to see the Theatre of the Tesseract through seeing in the Physical Theatre in Berkeley.

I got interested in the concept of time from literature and the philosophers; then when I had taken my degree in Physics at UT I got onto the idea of the modern mind being able to see into higher dimensions. I wanted to understand and make concrete through play in art, the abstract world view coming into our phenomenological époque. I was on the look-out for examples. I thought to create an amphictionic [1] theatre based on the exploration of one's tesseract. [2]

We are beings in space, as opposed to being on flatland. A flat lander of the labyrinths would only have left and right, or forward and backward in his world. But a being of 3 dimensions like ourselves, sees these and also sees up and down: he gets the extra dimension of depth. A 4 dimensional being would see length, breadth, depth AND the next dimension, which is time. Imagine a mouse running through a maze, it perceives only the corridors of its habitat from its point of view. If we were watching the mouse from overhead we would see all the passages of the maze and where the mouse would need to go to make an exit. Or how that mouse could just go over a wall. If we were watching ourselves watching the mouse in the maze, we would be like a 4 dimensional being who could perceive the whole history of the mouse as

---

[1] Amphictionic Theatre relates to the oracular pronouncements in ancient Greece

[2] A tesseract is the 4 D analog of the cube, which in turn is the 3D analog of the square.

corridors of time, a series of rooms lifted into an infinite hotel of all our labyrinths. And as we view and review the layers of passages of the maze and know where the exits up and out and sideways into time are, Proust and T.S. Eliot and J.W. Dune come to mind.

I was inspired to imagine Tesseract Theatre by the Personal Theatre in Berkeley and Montreal and in SF. I had videotaped some of the performers. When I say video taped, it was more like I had gotten into a dance with him, while he was performing, since his movement in the space was a dance, a drift, a long tai chi form through memories and feelings. Or abruptly he threw these faces at you; I used the video camera to receive these faces, I used the zoom to make the face move toward the viewer like they were coming down tunnels of space time, like they were part of the rooms of a tesseract. When I saw the actor of physical theatre perform, he flung me around on a psychedelic trip.

I followed him around. I studied Grotowski and Poor Theatre, and Artaud, and about working ensemble, in which the playwright takes notes on process. This is not production of product, but refinement of raw emotion, distillation of spirit. Process is the perpetual emotion food of this art.

In order to develop the outside eye to watch yourself work, the solo artist brings out the peripheral organizing system of his personal mythology. The unfolding and refolding of his time line. The poor actor on the stage with nothing, creates an atmosphere, a landscape in the mind, with gesture.

In order to understand, I created the solo theatre piece *The Subliminal Kid* from flashbacks and old photographs, the earliest memories and a dream. It is how a child learns to walk. It is infused with the physical exercises I had learned from studying Tai Chi.

The Subliminal Kid
—a solo performance piece for luminous theatre

Actor walks out onto bare stage. Spot cones down on him. He is wearing an obviously cheap suit, its shoulders are gigantically too big; it is some horrible shark skin. It has to be comical and uncomfortable looking. In spite of this he is trying to convey wealth despite lacking it. He starts talking to the audience.

>Hey, how's it goin', man?

(Indicating the suit.)

>You like the suit?
>I was at the St. Vincent de Paul Store, man, looking at the men's suits, and I come across this groovy looking 3-piece suit, I try it on and it fits.

(Shows label.)

>Got a label says hand tailored by Chinese tailors in Oakland! And Wow! It's marked DOWN! From $15 to $12.50.

(Mimes handing cash to clerk.)

>So I went up and paid the lady.
>And now I got this new suit on my back, which is very propitious, man, cause I'm on my way to apply for a JOB! But it's weird, man, cause whenever I put this suit on, I also seem to put on the persona of the previous owner, a type of guy who was, FOR SURE unappealing to my taste.

(Actor wriggles and writhes, tries to escape his own skin. Then looks pensive, wrinkles brow.)

>I can't tell if it used to belong to a dork, . . . or a slick. Or

maybe a slick dork. A dork who thought he was slick?

SHIFT— (Goes into lecturing mode.)

A slick is a kind of person, a kind of person—ality. . .

(Smooths back his hair with both hands.)

He's got slick hair, smooth, in place. He's an extrovert, kind of a controller, he looks at other people as a way . . .

(Makes eyebrows go up and down, looks imperious, haughty.)

of furthering his own ends. He's . . . concerned with appearances,

(Waves hands as if touching a smooth table surface.)

and the surface of things. . . A dork on the other hand,

(Actor ruffles his hair, shows teeth in an overbite.)

is an introvert.

(Brings shoulders up to ears, looks diminutive.)

He is usually the one being controlled. He may be more sincere but spends too much time beneath the surface of things and can't relate, or is so afraid, he can't share.

(Actor addresses audience directly.) Then SWITCH —

On the other hand, the Slick is so saturated in image he feels like he has to fill the emptiness in his life with image and random sex. You've seen them at singles bars. Being into the scene instead of the people, looking at themselves in the mirror, constantly on the make for sex.

Slick: "Hi! Can a guy, buy, you a drink?"

Slick men trying to pick up slick women. He's tough-minded, she's articulate. She's glitzy, he's glib. He's upbeat, she's glamorous. He's superficial, she's narcissistic. Like in the fashion photos not showing any real emotion.

Tele-cocks grinding against video-cunts a go-go.

The slick want's to be famous without doing anything great.

Actor makes an "on the other hand" gesture. SWITCH —

On the other hand, the dork is caught up in a never-ending struggle. Constantly striving and never having any fun.

"Dork: 'Just drink more milk. And work harder.'"

The dork is serious, overly concerned with his career, a type of guy who would NEVER be unemployed or be satisfied as a blue collar worker.

Actor rubs his own arms in the suit, addresses audience. SWITCH —

I don't know, man, I guess I must identify with him somewhat.

Whatever he was, he was STRAIGHT.

(Wraps arms around himself, hugs, holds himself tight.)

I feel this suit envelope me like a straight jacket of guilt, I feel the dork in it wanting to be happier, more into partying and hedonism, and the slick wanting to have more depth and responsibility.

SHIFT—. Break Frame. Actor walks away, turns, address audience.

I need this suit 'cause I'm looking for a job. I am sweating like the slave of an evil spirit. A flat broke hippie with huge gaps in my resume . . . vast panoramic gaps, man, so big an elephant could waltz through, like it was a shift in time, in which I hitchhiked around the country checking into things.

(Shakes head in resignation)

>Nothing but the blues, man.

(Looks thoughtful.)

>Sometimes I regret — not having that certain —
>"friendly," blood thirsty, attitude

(Looks maniacal, aggressive.)

>it takes to "stack chips"
>to begin rising — to have power —over a dip-shit middle
>class — brute class. To become prominent — enough — to
>have to, I don't know,

(Shrugs; looks doubtful.)

>to be assas-in-ated in style.
>But I have to do it, cause I was looking . . . for a . . . job.
>
>There I was,

(Actor strikes poses to illustrate the following types.)

>imaginary prize fighter
>imaginary saint,
>imaginary gigolo,
>spiritual athlete . . . complete
>with teflon vest and bullet proof limousine.

(Lifts hands in resignation.)

>So there I was, an imaginary dual status alien, with a
>social insurance number   from the country of the
>unemployed,

Actor is looking up at something on a wall . . .
>standing in the foyer of the Consolidated Building, man,

reading aloud the names of the firms from the directory.

(Reads from the placard on the wall.)

>ITT, Summa Corp. Texas Instruments, Eckancar, General Motors, Texaco, the Catholic Church, Enco. On the 16th floor a firm called Geometrical Optical Dynamics.

Actor Whistles, raises eyebrows, spells out the letters.

>G _ O _ D   My God! Fantastic!

Actor does the Fair Lady works at Shuttles movement, sweeping upward into a corner,

>And I go up 16 floors at once, man, hesitate outside the door marked Geometrical Optical Dynamics then go in.
>
>An attractive secretary comes down the hall, she is wearing polyester print, hose. She has frizzy hair and gooey eyes.
>
>I follow her as she takes my resume and leads me in.
>
>She leads me into the office of a stout Indian man. He is short, squat and powerfully build, well tanned and bald. His name is Mr. Yaqui. He looked at my resume in front of him.

SWITCH — Actor plays both parts of two character dialog — Mr. Yaqui and Applicant. Mr. Yaqui looking down at resume and up at applicant.

Mr. Yaqui: I see here that you have some work in a sleep shop.

(Applicant looking down at Mr. Yaqui.)

Applicant: Yes, I awakened dreamers 3 times a night, and
           asked them to divulge the contents of their dreams.
           I worked on op-amps and other medical electronics.

(Mr. Yaqui gestures — Follow me. Leads applicant down the hall looking over his shoulder as applicant.)

Mr. Yaqui: Yes, well let me show you our latest product. We call it the microprocessor of dreams.

(SWITCH — Actor is narrator. Explains what we are seeing.)

He is holding what looks like a bicycle helmet with a lot of wires coming out of it. The wires are attached on little round golden spots on the smooth helmet surface.

SWITCH — Actor as Mr. Yaqui. explains.

Mr. Yaqui: "It extends the idea of voice recognition by a computer to brain wave recognition by computer over a wireless link from all these sensors in the helmet to that computer over there.

(Indicates something across the room.)

Mr. Yaqui: For every word there is a characteristic waveshape associated with verbal part of the cerebral cortex.

(Indicates left parietal lobe of the brain.)

By means of transducers attached to the head   here. These waveshapes are transmitted by F.M. modulation, to our computer. Not only that but other centers of the brain have characteristic information transduced and transmitted as well.

(Mr. Yaqui points to something he is holding in his hand.)

Mr. Yaqui: Sensors slightly offset over the eyes pick up the landscape of Rapid Eye Movement — *(pause)*

(Actor switches to Narrator, does exaggerated Balinese Eye Sweeps)
(Actor switches to Mr. Yaqui)

Mr. Yaqui: — and construct the image graphically.

Other sensors pick up visual perception,
volition and abstract imagination; these become direct
sources of data and command information
for the Dream Machine
(Holds up the helmet)

Mr. Yaqui: We call this transducer device, that goes around the
head — Shiva's Headband.

Actor looks afraid
SHIFT —: Actor as Mr. Yaqui is assuring, encouraging.

Mr. Yaqui: The currents are tiny, on the order of nanoamps.
With these small currents, we can run all transducers
backwards — transmitting directly back into the body,
the amplified and clarified signal –
enabling the brain to absorb signals from the machine,
INDEPENDENT of normal sensory channels.
It's like riding a bicycle.
The dream machine is a super-computing extension
of the operator's own natural abilities.
The feedback it provides facilitates evoking
a direct perceptual insight.

(Mr. Yaqui addresses subject directly, performing hand gestures
suggested by story — a gyroscope, a vortex to enter the mind)

Mr. Yaqui: But here, I could describe the dynamics of riding
a bicycle with a complex matrix of simultaneous
differential equations, relating speed, curve of path,
weight, whose solutions are spherical harmonics
related to the gyroscope problem on a moving frame of
reference.
But a kid simply jumps on the thing
and *feels* the right thing to do.
And so can you.
Be the operator of the dream machine, just *feel* and *see*
and *steer* your way.

SWITCH — back to Actor as narrator

> So next I was in this cubicle adjacent to the machine room. And Mr. Yaqui adjusts the headband over me. Calibrates it. Then switches it on!

QUE: Colored gells change scene lighting. Speaker in dark. The actor's voice is picked up, processed with cavernous reverb.

> An eerie sensation seemed to take possession of my mind, man.

Actor begins spinning like a dervish, in and out of the colored lights and shadows.

> It was like waking up
> inside a huge parabolic glass bell,
> and shouting at the top of my voice
> and hearing only a whisper as the sound was reflected away,
> but now these were my thoughts snatched away, man,
> and then they would come tumbling back.

Actor is stopped in a spot light. He is recognizing something seen, or remembered. He staggers momentarily, realizing it. He tells a memory. He takes us into the body of a dream.

> I am joy riding in a stolen car. Driving around in San Antonio. Near Jefferson High. Out into the sparse unfinished housing estates.
> Among the vague terrains, on the outskirts of the city.
> Me at about age nineteen, and the son of the owner of the car, who is tall lanky, gawky and looked like me. He is some kind of alter-ego of my current personality. In fact he is a younger version of myself.
> Then I realize: He's the Subliminal Kid!

Actor walks around an invisible object like he is looking at a statue from all sides. Then he steps into the spot light and becomes the statue. Then begins watching an imaginary observer walking around HIM!

Him and me end up driving around my old high school, man, looking for a space to park it. We drive up endless rows of boulevards that writhe like some kind of labyrinth, or like Braille ideograms of the DREAM, man. After parking the car, we went up some dead end street and had to come back, and there were the police looking all around the car.

We were dressed casual and look like very young boys. We looked out of place. And HE, wanted to GO BACK to the car!

I told him to keep coming with ME.

Actor switches off playing the Kid and the Narrator.

The Kid: "But I think I should go back to the car."

Actor: I tell him, "Naw, come with me."
    And he does, and we set off on a jog,
    and he followed me down along the river. — *(pause)*
    From there the dream turned into one of those endless dreams where you are trying to go somewhere like in this space,

(Actor looks at audience, makes a sweep of the hand gesture equivocating space in story to this theatre space.)

And you just can't get there. You keep moving down endless side streets in this labyrinth. The streets of the suburbs with their vast yawning lawns and trees and shrubberies. I am feeling out of place because I am dressed sloppily, like a slacker, like a big kid.

SHIFT — the following is performed in sometimes graceful lilting dance movement like a Viennese waltz, and sometime like a martial artist going around the Hsing I Pa Qua circle doing martial arts moves. The images is like waves going around a circle.

    Endless labyrinth: . . . I . . .

STREAM — *(Actor reaching out into something, feeling in the dark. Like climber trying to pull himself or reeling in feelings. Like a blind man feeling his way through a vague terrain.)*

> feeling . . . touching . . . branches of feelings —
> the eyes receive and retransmit.

(Said like making a discovery.)

> The light fills the mind with clarity
> and the light empties back out onto a projection.
> The sun enters and penetrates through
> and around us like the wind,

(Actor makes a brushing off into the distance hand gesture.)

> and the eyes sail . . . on this wind.

Actor makes exaggerated movements of the eyes, like Balinese dancers
> And there is a space of Rapid Eye Movement
> in which the eyes are in continual movement
> and this scanning space
> of subliminal video games occurs.

(Said like seen for the first time ever right here in front of you.)

> And it is an immense space.

Actor sinks low, looks up

> At the bottom of the sky . . .you can see yourself supported in the ball works of struts extending down from the atmosphere of clouds into the water worlds of the hydrosphere and the geosphere of the land, things falling down and being given. We are in a space-time diamond.

(Demarcates a closed cubical space structure with his hands, Fair Lady at Shuttles)

> a 4 dimensional cube or tesseract. . .

Actor turns and pushes away behind him

> where The Past is behind,
> but also impinging on the present.
> And one can stand off and float as above and see one's self from above . . .
> See more in the air this way than in a mirror.

(indicates the theatre space as though it were physical)

> For this is your mirror.

(convey a sense of surges of force coursing through)

> The air has a depth that can buoy you up,
> as if we could fly in *this* sky.

Actor spreads arms out like wings flying

> On feelings. . . as they are always there . . .

Actor mimes the positioning prepositions in the text:

> . . . behind the hands and just out of reach
> . . . behind the eyes and in front of us
> and getting . . . between us . . . and what we do.
> And they are trying to advance towards us,
> so we can see them
> but can't quite get our hands on them.
> They are always receding like the horizon.

Actor looks pensive addresses audience as they are his eternal child archetype

> What does it mean?
> Who is this alternative me? That I am dragging along.
> He is the innocent child, who can't be allowed to feel because of how he had to be.
> I am the Rebellious Child.

And he is the Innocent Child,
the good child, loved simply for being?

(Thoughtfully coming to this realization.)

It means that the feeling child wants to be up with me, the older one, but is drawn back to the authority of the parents, drawn back to all kinds of authority. And that without each other, we are lost, doomed to run in an endless labyrinth.

SHIFT — Actor gestures release, something falling away. He addresses audience, softly.

They're like . . . so many masques to you . . .
And they just fall off. . .

(Argumentative, accusatory. Like a child teasing.)

You see through them. I know you do.

Actor looks contrite, rueful

Oh . . .. sometime you let me charm you.
Sometimes you let me think I can make you laugh.

Actor looks pensive realizes. Makes argument with the absurd:

It hasn't occurred to me until now,
that with each new mask . . . I portray to myself . . .
the previous one dissipated into death;
which is a kind of life.

(thoughtfully)

Non-existence would be a more accurate reference
to the old mask . . . who was so humble.

Actor starts to get very agitated, wound up, hyper. Grandly:

> Entrancing with his gaze,
> anyone who'd follow him, into his maze.
> In the center of their labyrinth, would be a secret place,

Actor from the center of the stage speaks tentatively, as though he were trying on the sounds of the words or shouting them into an echo chamber

> they would call . . . their True Love.

Actor apostrophizes, speaking directly, breathlessly to the shadows at the edge of the stage

> My love . . . my true love,
> she is the self — same as I am — we
> were always on the lam
> via the image... in nation.

Actor looks puzzled, quizzical at the audience

> Or perhaps people are the states.
> Or perhaps not even people . . . say, individual
> personality is the state,
> is the means. And say it is in face possible for an individual,
> an in-de-viz-u-al,
> to posses a variety of states
> — a nation within themselves United
> the United States of Hysteria.

STREAM —: Actor looks vulnerable, slightly camp, apostrophizes dancing in a pas de deux with shadows of the space

> Dearest true love . . .
> you are as free as a state of this hysteria
> as I am free . . .
>   free...
> and as my only concern
>   is a balance...

Actor reacting as though being laughed at:

> You laugh . . .
> > with your eyes . . . and ask . . . openly
> > aloud . . . what is this balance
> of which ye . . . babble, my love?

Actor driving home a point

> > I say it is a balance of possession
> >   and . . . of total freedom.
> > You understand this already.
> > I know you do . . . and as we are free
> > we are free . . . to be . . . possessed
> >   by each other.
> > This is the choice . . . this is the
> >   balance   of freedom.
> > Now the obsession
> >   is a loss of equilibrium
> > the loss of the free will . . .
> >   Could it be . . . the meaning
> > of falling . . . in love?

Actor resolves into a kind of relaxed open stance.

SHIFT —: Becomes more agitated. The visions of the experiment are amplifying, he is seeing more directly into his feelings.

> I am struggling to wake up!

Actor mimes and makes his face into a masque to illustrate the following states

> > I come here . . . I laugh . . . I cry...
> > I stare like a child . . . I relive
> >   those feelings
> > I had when I was a child . . . the terror . . . the joy . . .
> > I forget . . . I pretend I don't understand
> > I guffaw, I write

> I hugamugga!

Actor looks resolute, dejected
Actor swirls head around in a gesture of haplessness, giving up

> Death has such a big Mouth,

Actor pulls back inside himself

> I should be quiet and live in peace.

STREAM —: Actor swirls head around in a gesture of haplessness, giving up, and frightened away

> Death has such a big Bloody nose,
> seamless all white clothes,
> And endless number rows of TEETH
> in my theatre

Actor sweeps hand indicating space

> for this is the theatre . . . in which I dwell.

Actor sweeps hands down indicating his body; heavy shouting emphasis on sell

> In the body . . . which I must not sell!

SHIFT —: Actor looks rueful, like he has been taking himself too seriously. Smiles. Address the audience directly.

> With tales like these, who needs a head!?
> —Vietnam mom
> *(sings)*
>
> My little babe been to Vietnam
> Got his head blown of but resewn on.
> Resewn on, Resewn on,
> Got his head Blown off, but resewn on.

(Says)

— They got some mighty fine doctors in Vietnam.

SWITCH — Actor goes back to narrator telling story of job interview, aftermath.

And then suddenly it was gone.

(Gestures lifting a helmet off the head)

I felt Mr. Yaqui remove the headband.
And he was looking at me and grinning.

SWITCH — Actor as Mr. Yaqui, looking at narrator

Mr. Yaqui: It's okay; I just switched it off.
Did that blow your mind?
Everybody gets that the first time.
See, the Dream Machine acts like
a gigantic feedback loop for mental processes.
You can get into a positive feedback situation, where anything floating around in your mind, gets . . . gets precisely defined,— quantitative.
Then sent back to you, amplified!
You give it junk, and get back super junk. Then take the super-junk, and think of it, and get back super-super junk!

SWITCH — Actor is narrator telling story

And then I'm back on the street again, man, but it's not the same. So many people, so many cars, so many stores, so many bars. Windows wink down dazzling me with solar reflection. And that shining — off the chromium cars — fries my eyes.
Everybody else knows exactly what they are doing, every minute. Everybody walking fast and talking fast,
pouring out of restaurants and stores, sweeping around corners, surging across intersections, up stairs, into buildings.

I am swept along, man, into the impossibly, crowded,
subway.

Actor has hand up holding subway car strap, looking with fear at crowded subway riders. Indicates he is completely jammed up all around.

Thousands of beings all around me packed, stuffed,
wedged, and there is absolutely no air, man — to breath.

STREAM —: Actor gets very agitated wound up, hyper. Hits rhythm of list like a train down the track.

Riding riding, riding
22nd avenue 32nd avenue
Waitress with vampire eyes — wanting to spend the day hanging
upside down from the overhead railing.
Her feet are killing her.
42nd avenue   52nd avenue
riding, riding 50,000 hate vibes par second
being given off   from the biomass.
We drive through a cloud of what seems to be bean fart.
What heavenly majesty!
62nd avenue 72nd avenue
Paranoia rising. Anxiety rising. Hate vibes.
Intense fiendish energies,
collect down in the tunnel
Death! Kill . . . somebody!
Cut somebody! The moon made me do it!
Push in front of subway.
Jump yourself!

Actor gets shouting louder and louder

82nd avenue   92nd avenue
Fart   Sweat!   Foul!
Arrests   Narks!   Busts!
Ritualistic sex crimes! Cattle mutilation!

Actor shows long-suffering face; trapped

> 102nd avenue
> I am wedged so tight in that I can't get off.
> Turn eyes to newspaper to have something to occupy them.
> Reading headlines: Psychosis plagues. Contagious schizophrenic virus unleashed by irate genetic researcher.
> Lawyers, undertakers, doctors doing   BOOMING business.

Actor getting worked up, rising to shouting:

> Plunging down the tubes into dark night, man.
> Bat with baby face goes WOOosh by my head.
> I go into mantra to stop fear:

SHIFT —: Actor pauses in silence. Looks toward heaven. Sings. Song ends with mouth hanging open.

> Sometimes . . . I like to let my mouth hang open . . . and look dumb.
> Sometimes . . . I just like to let . . . my mouth . . . hang open. Just to try . . . to look . . . dumb.   Just to try . . . to feel . . . a little dumber.

SHIFT Actor returns to being narrator

> Finally I'm back, to my pad, my little cave to crawl into and fight back from. I am talking to my dog, man.

Actor bends down addressing dog, teasing.

> How about . . . going . . . for a walk? . . . in the park?
> And we are out!
> Walking beneath the freeway, man, down the hill past the heavily graded train track, man. Me, and Sunshine the pit bull, walking through the tall grass past the freeway construction, across the bridge, over the river.

Actor looking entranced at a shimmer of light

    I love to watch the green water in motion. I lean against the rail and look out: I wonder about where the river leads to. And wish I was on it. I look over the edge, watch the pattern of lights in the undulating water, and think about how all matter is waves of energy held together by various forces of attraction.
    AHHHH, it's a lonely life man.
    Breathe in a little O_2 Exhale a little CO_2

(shrugs in acceptance)

    Still there aren't that many forces of attraction on me.

(looks thoughtful, talkative)

    Lately I've become religious about atoms. Every night Walter Cronkite comes on T.V. with a new cause of teenage cancer.

(winks at the audience, smiles)

    They're found that you can catch—on fire—from a toilet seat.

SHIFT —: *(Back to being serious)*

    I worry about getting my b_{12} complex.
    Worry about keeping a proper balance in my metabolic ecology.
    Will the bad wigglies overtake the good wigglies?
    Am I getting enough chromium?
    What about the polarity of my potassium?
    This is how I have come to dig the element: Man —
    freak with 103 personalities — the do it yourself chemistry set.

(Rotates around in a dervish spin (arms out the Feynman diagram))

    In the spin of color, charm,   beauty   magic & strangeness
    there is a principle that penetrates people.
    It is a subtle body, invisible and untouchable

(indicates the directions in the following)

>which circulates energy from the earth to the sky.
>Through and up and out and it is the combined spectrum
>>of the elements in a being
>>>— for we are radiant beings
>>>>and this fire
>>>in the cells of the body
>>>>in which I dwell
>>and in which . . . all the stars and planets
>>in the universe are convolved,
>>sometimes shoots out wild sparks in the dark.
>>They trail off and are never heard from again.

(indicates the horizontal with sweep of hand)

>And the line of the horizon is the edge of the eye,
>and the earth is an immense eye in the face of space
>a space which no being can see.

(spreads arms in chest expanding gesture of breathing then circulation)

>A kind of inspiration and expiration of the firmament
>>in our breathing,
>>and it surrounds us and travels through us,
>>all through our limbs and makes us tremble like
>little antennae or cilia on the tiny water creatures . . .

Actor looks punkish, contrite taking leave of the audience shrugging it off

>Still . . . there aren't that many forces of attraction
>on me.

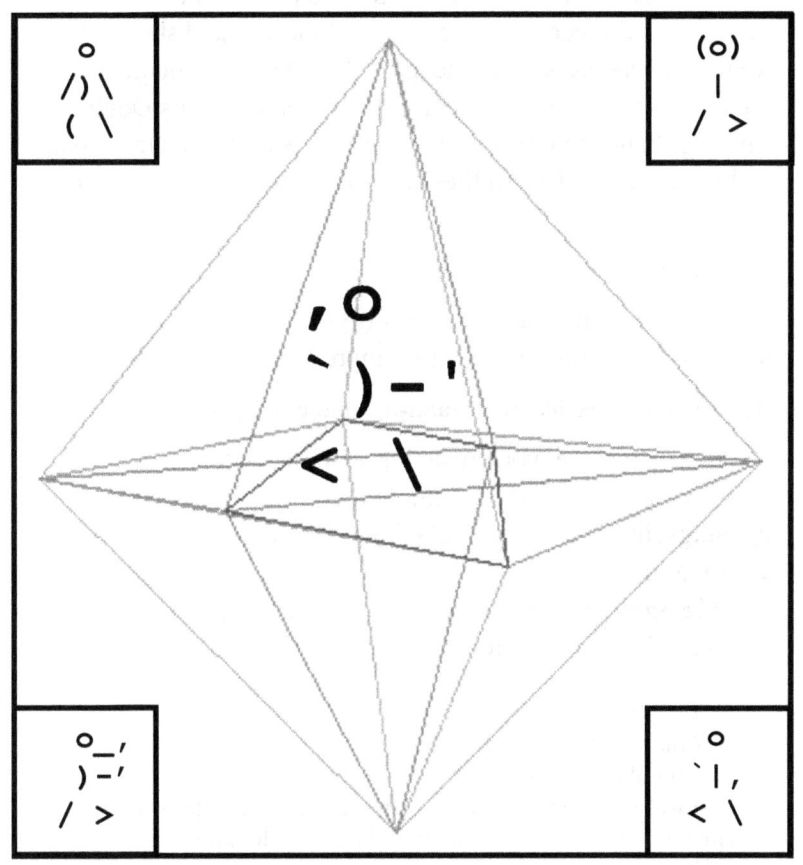

The Theatre of the Tesseract (notes)

The Subliminal Kid is a solo theatre piece in the tradition of Grotowski's Poor Theatre, physical theatre, and stand-up routine. There are also elements of sacred drama, mudra theatre. It is inspired by Artaud — the theatre and its Double. The big change that this writing art brings to the creator is that it shifts the focus from a literature of content to a literature of process.

The Solo actor plays many parts.

In this minimalist theatre of storytelling, you have to be tech, writer, performer, director, impresario and dramaturge.

Address audience like in a stand-up comic routine.

In a solo piece, you are acting with the audience as your scene partner but they don't know it yet. You have to leave openings, invite them in. Make it easy for them to come in to your world.

The Subliminal Kid begins with the most open form: the (hapless) stand-up comic.

> Hey, how's it goin', man?
> *(indicating the suit).*
> You like the suit?
> I was at the St. Vincent de Paul Store, man, looking at the men's suits, and I come across this groovy looking 3-piece suit, I try it on and it fits. *(TSK 241)*

It is wise to start with presenting yourself as lower in a hierarchy — scared, even. Respectful. You don't know what the audience will be like; you have to hook them into your conversation, into your world. This leads to intimacy.

Solo actors bounce their work off an unpredictable crowd,

they are not fellow actors whose performances are more or less the same each night. The solo theatre piece is portable, it must appear to be a spontaneous rap. And it must be adjustable, to fit the audience's mood.

You are going to take them inside your life through your talking and gesture and imagery, and it is this immediacy, this sense of interacting with the moment in the room that is most prized. So be committed to that openness. It is disarming.

Solo actor uses a line of symmetry — crossing the line to inhabit, impersonate characters. SWITCH —, SHIFT— and STREAM.

The solo actor who is going to play more than one character has to help the audience see the "*SWITCH —*". I use "*SWITCH —*", to change characters, and "*SHIFT—*", to change the focus within a character's story.

> *Actor makes an "on the other hand" gesture.*
> On the other hand, the dork is caught up in a never-ending struggle. Constantly striving and never having any fun.
> "Dork: 'Just drink more milk. And work harder.'"
> The dork is serious, overly concerned with his career, a type of guy who would NEVER be unemployed or be satisfied as a blue collar worker. *(TSK 243)*

Camera = World x View x Projection.

Think of it as Camera = World x View x Projection.

You as writer have to indicate how close-in the viewer is invited, even required to be. You are telling the viewer where to look and how closely to inspect the imagery of his own mind, that you have led him to view, with the imagery from your own mind. Since you are acting in 3-dimensional space, you have more dimensions to project onto the screen inside the viewer's mind. (See these point of view shifts in the chorus of the poem, Tesseract.) (ITN 179) It is like you have to be the lens of a camera: where you look, he will look. Where

you zoom, he will zoom.

There is a drama in this interactive tripping together.

The story draws from personal experience.

I worked in a sleep lab at UT, and we were tasked with maintaining the equipment. They awoke sleepers during the night and had them divulge the contents of their dreams. So this got into the story of the Sumbliminal Kid.

> *SWITCH— Actor plays both parts of two character dialog. My Yaqui looking down at resume and up at applicant. Applicant looking down at Mr. Yaqui.*
> Mr. Yaqui: I see here that you have some work in a sleep shop.
>
> Applicant: Yes, I awakened dreamers   3 times a night, and asked them to divulge the contents of their dreams.
>
> *(Mr. Yaqui leads applicant down the hall.)*
> Mr. Yaqui: Yes, well let me show you our latest product. We call it the microprocessor of dreams. *(TSK 245)*

This technical talk that Mr. Yaqui gets into, reflects the metaphor of Jung's "Archetypal Regulator" which is like the Totemic Operator of Levi- Strauss. (See picture (at right) and on page 277.)

Entering the body of the play.

In the following passage we are further entering. We entered the former owners of the suit. We entered the office building in search of a job. Now we are entering the dream space through the dream machine. The character, Mr. Yaqui is speaking. He is the don Juan of this anthropological exploration into a scientific paradigm which is running parallel to the modern citizen's quest for understanding his own purpose and abilities.

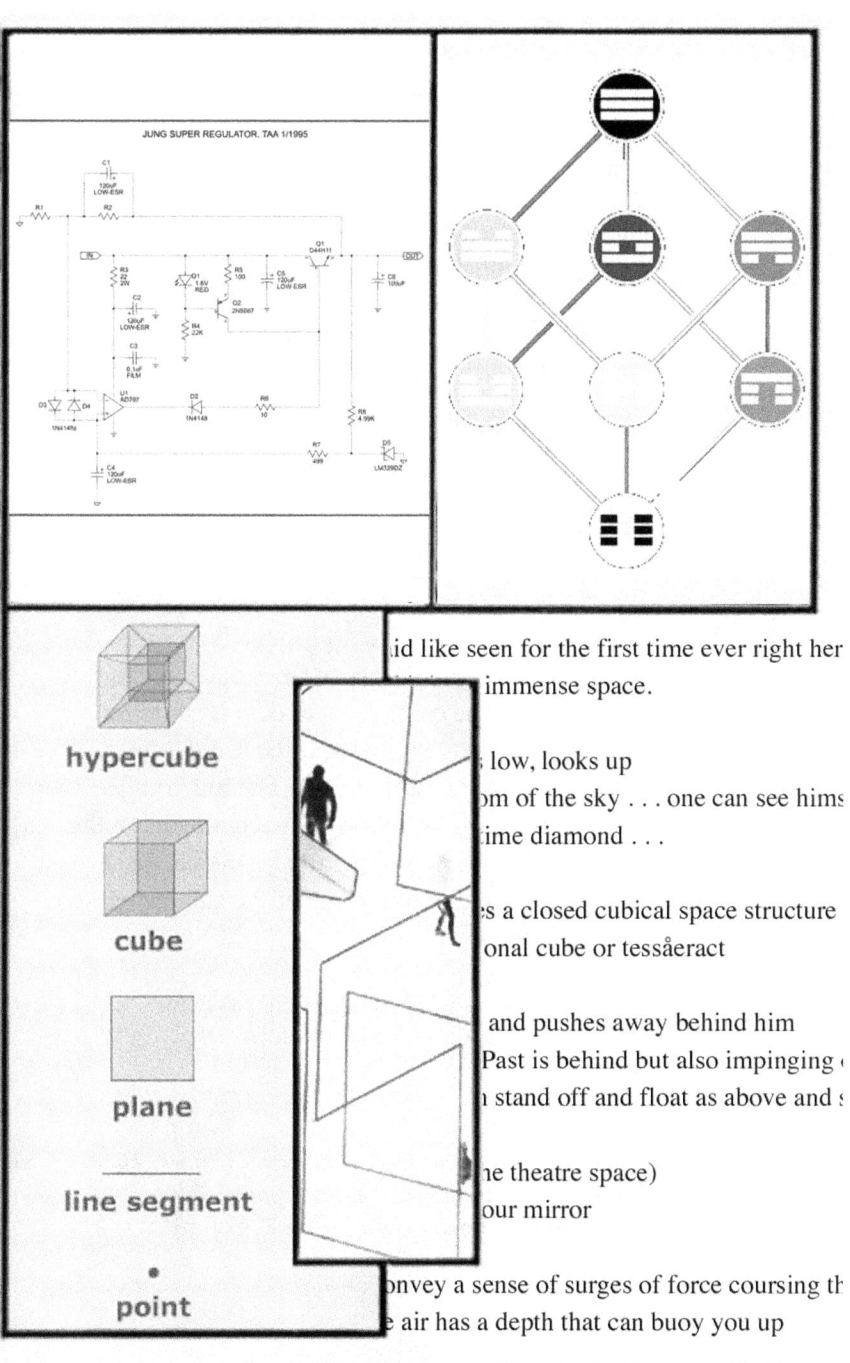

...id like seen for the first time ever right her...
...immense space.

hypercube

...low, looks up
...om of the sky . . . one can see hims...
...ime diamond . . .

cube

...s a closed cubical space structure
...onal cube or tesseract

...and pushes away behind him
...Past is behind but also impinging...
...stand off and float as above and s...

plane

...he theatre space)
...our mirror

line segment

...onvey a sense of surges of force coursing th...
...e air has a depth that can buoy you up

point

Mr. Yaqui: The currents are tiny, on the order of nanoamps.
> With these small currents, we can run all transducers
> backwards —transmitting directly back into the body,
> the amplified and clarified signal –
> enabling the brain to absorb signals from the machine,
> INDEPENDENT of normal sensory channels.
> It's like riding a bicycle.
> The dream machine is a super-computing extension
> of the operator's own natural abilities.
> The feedback it provides facilitates evoking
> a direct perceptual insight. *(TSK 247)*

Drama.

The script is a blueprint for a performance.

Most of the time in writing plays, we tend to explain everything. Well, that is not alive, that's not urgent. Drama is here and now in the moment. This art is a theatre of the intuition — the art of implying things. It takes place on an internal stage. We have to be careful not to devolve into too much technical specialist language. But at the same time we don't want to be dumbing things down. The hip young audience of today is very sophisticated. It is incumbent on the artist to play with all the new ideas making up the zeitgeist.

Be dramatic, to ensure that the image of the subject being hooked up to a machine, is conveyed.

> So next I was in this cubicle adjacent to the machine
> room . . . and Mr. Yaqui adjusts the headband over me.
> Calibrates it.
> Then switches it on!

QUE: Colored gells change scene lighting. Speaker in dark. The actors'v voice is picked up, processed with cavernous reverb.
> An eerie sensation seemed to take possession of my
> mind, man.

Actor begins spinning like a dervish in and out of the colored lights and shadows

> It was like waking up
> inside a huge parabolic glass bell, *(TSK 248)*

A real dream is told.

This theatre is an art of empathy. When people see that the character before them is also the person, there is an empathic resonance of identities. This intimate theatre based on the rap, or the routine, or the lyrical rhapsodic ode, is a journey of exultation and profanation that the audience is brought on, through this resonance of identification in the real. Vulnerability can be powerful if you can disengage and shift around among your sub-personalities or channels or centers. It takes a lifetime to become this spiritual athlete. Though this retelling of a dream is perhaps not as dramatic as the other activity in the play, we are going for a moment of truth here in the art. This takes the pace down a bit getting ready for the next rise.

> *Actor is stopped in a spot light to tell a memory*

> I am joy riding in a stolen car. Driving around in San Antonio. Near Jefferson High. Out into the sparse unfinished housing estates.
> Among the vague terrains, on the outskirts of the city.
> Me at about age nineteen, and the son of the owner of the car, who is tall lanky, gawky and looked like me. He is some kind of alter-ego of my current personality. In fact he is a younger version of myself.
> Then I realize: He's the Subliminal Kid! *(TSK 248)*

The solo play moves into a dance of physical theatre.

> *Actor makes exaggerated movements of the eyes, like Balinese dancers*

> and there is a space of Rapid Eye Movement
> in which the eyes are in continual movement
> and this scanning space
> of subliminal video games occurs
>
> *(Said like seen for the first time ever right here in front of you)*
> and it is an immense space.
>
> Actor sinks low, looks up
> At the bottom of the sky . . . one can see himself
> In a space-time diamond . . .
>
> *(Demarcates a closed cubical space structure with his hands, Fair Lady at Shuttles)*
> a 4 dimensional cube or tesseract *(TSK 250)*

The Fair Lady Works at Shuttles movement in Tai Chi, is a mini-mudra in which the player acknowledges the 8 directions of the Pa Kua. (See page 264.) This can also be seen as a tesseract as we will show in the later half of this essay.

I will just mention here the play of Faces, Edges and Corners that is the Tesseract. (See page 271.)

As suggested in the Pa Kua, the physical theatre becomes metaphysical

The solo theatre piece here starts to become ironically self-aware as the actor does a riff on masks.

> *Actor looks pensive realizes*
> It hasn't occurred to me until now that with each new mask . . . I portray to myself
> the previous one dissipated into death;
> which is a kind of life
>
> *(thoughtfully)*

> non-existence would be a more accurate reference
> to the old mask . . . who was so humble; *(TSK 252)*

This physical trope sets up the opportunity for lyrical rif.

Actor gets very agitated wound up, hyper — grandly

> entrancing with his gaze
> anyone who'd follow him, into his maze
> in the center of their labyrinth would be a secret place

Actor from the center of the stage speaks tentatively, as though he were trying on the sounds of the words or shouting them into an echo chamber

> they would call   their true love

Actor apostrophizes, speaking directly, breathlessly to the shadows at the edge of the stage

> my love . . . my true love
> she is the self — same as I am — we
> were always on the lam
> via the image… in nation *(TSK 253)*

Discussion of the model and aesthetics

The Subliminal Kid was a ritual to pay one's dues to the existential commitment undertaken in the plunge into your own time. Live, as art. I had to swallow my fear and go on. Walking out onto the stage, the state space, with the big suit, I stepped into the character. I was swallowed by this Ambivalent Man character, the quintessential wishy washy hippie. For better or for worse, that was me in those days. He was sure of himself, cool. He takes his time, doesn't jump right into it, makes them wait, wets the appetite. For we are here in theatre, to explore the processes of savoring being.

If you don't have much experience in acting, playrighting,

and show business, then you must rely on process, critical technique, and abstraction to carry you. Get grounded in the scene: We are at some kind of comedy show, he is an entertainer up there doing a routine. He is coming into a kind of self-reflexive being through charming his way into the minds of the audience. We are attendant to birth, accompanied by the coincidental cries of the world outside: the wail of a siren in the night, the sound of a passing car in the street, or the wind rattling the doors and window panes of the theatre.

The comedian steps into the vortex spotlight. He is in the presence of his own truth. He has practiced so much, he has the stamina of a spiritual athlete. And he can exert such control that he can do anything with his body, enter states of joy and ecstasy. He can rotate into memory. He draws the viewer in by being submissive, telling a self-deprecating story about his poverty and trying to rise above it, as he tries to act excited about a suit.

Face-masks

Be able to make masks with your face. The actor is laying out the structure of the story; the backstory is a job interview that becomes a self-interview with his own truth. This gets him in touch with his inner child. All the while by gesture and eye telegraphing, he is creating space within his motion for the story to unfold. Let the movement change: tight and constricted, then blown out and flowing. To be able to make masks with your face undermines the pain and loss that starts to take hold of your physique. The solo artist protects himself by creating masks. The actor of the Subliminal Kid says: ". . with each new mask / I portray to myself . . ."

One mask is the giant personage to whom the subliminal kid is applying for a job. Distortion as seen from a child looking into the world of the giants, as the applicant and employee exchange dialog. The actor shows diminutive Appli-

cant and aggrandized Employer stature in the size of the stance of his frame. The big stuffed sport coat is gone now. He also depicts the employer speaking a technical jargon about the mind and computers and their interface.

The actor goes into a riff on Faces, when he has the brain amplifier attached. He does this reaching out and touching, touching his space, like a blind man feeling for clues, touching his body like he is in something, a room with shifting walls upon which shadows are being projected. Imagine the shadows as archetypal or parental Blake figures, of child looking up, and seeing the wall of Plato's cave and understanding. Or Goya figures in the dark malevolent forces moving through dark night. This yoking together of discrete organizational entities, sometimes opposites, is structuralist writing. We are inhabiting the armature of a body. We gestalt with the energetic shimmer of binary structures brought from opposition into proposition, and this melds binary categories.

line : square :: square : cube :: cube : tesseract

Faces — personae; corners — mask; edges —transition.

Faces — personae; corners — mask; edges —hard edge martial art movement. Corners: rotation. Here is where I got the idea of the tesseract. Mutually perpendicular activities, rotation in time, go around a categorical corner to become some other part of the psyche. Tesseract Theatre, the gift of a little cube in the hand. (Little cues from the indicating hand.)

The Actor lets go of one scene, turns to go around a corner by rotating on the stage in front of our eyes to another space. The way he acts is in the hard and soft edge of his movement (this is the demonic shadow). He can keep going around corners to past activity of his ego. Faces – personae; edges — shadow, corner—ego. The three in one of the other. The actor of the Amphictionic Theatre takes you to a place where defeat doesn't mean anything anymore. It's not poetry, it's subjective truth.

To me it was a kind personal storytelling. This should help one memorize the parts of the play with ease, because they correspond to parts of your own psyche. The dialogue of the play is a conversation between Ego and Id and various Personas in a mnemonic theatre.

In structuralist writing you are creating a vehicle to get into, a structure that resonates with the structure of the Other, how the other structures his thinking.

In the scene of the job applicant, the subject is trying to enter a hierarchy. Hierarchies are like family. Father is to Son as Manager is to Employee.

Hierarchical is to heterarchical as planned out is to experimental. Being in control is to learning as top-down is to bottom-up. And there is a female in it who does the human resources work. It is like the husband and wife. Father husband manager and mother wife HR-person are to son, child, employee.

We have these other entities in the play. The older self becomes the protector mentor of the younger self. Mentor is to mentee as powerful ally is to ego as uncle is to nephew. So we see the parallel scene of the employer is to job-seeker being played out again inside the actors own mind in the way the older self is keeping an eye out for the younger self.

The ordering of these parts of the psyche into a unity, into a whole, is like the contemplation of a mandala. While walking in the improvisational labyrinth you are in the mandala of theatre. So on any given day, the amphictionic actor could dwell in the functions of psyche. I might address the part with what's on my superego today, or what's on my id today. You must be like a boxer — in shape to go several rounds, for you are boxing with your shadow, and must come up with whatever it takes, to bring mind down to coincide with whatever emotional space body is in.

This is resolution.

Structuralist writing is like a crossword puzzle: the diachronic across and the synchronic down. This is the instrument that TS Eliot and Pound played, a weave of mythological elements across different cultures and times. This is the poetry of mythopoeisis. Myth as syntax or algebra is a matrix with elements across and elements down, generating the text. The text is a projection out of the matrix, out of the multidimensional database with attributes undergoing joins and mergings. The perceptual matrix is central to Eastern thought. Their concept of number is an embedded element of matrix. The I Ching is a great example of a mythological organizational structure.[1]

I was trying to take the things Pound was saying even further. I could see in his sculptural concrete poetry, that he had done for the marriage of words and sounds what Descartes had done for the union of algebra and geometry. In his essay on Vorticism, he reflects on the progressive intensities of mathematical expression, namely: the arithmetical, the algebraic, the geometrical, and that of analytical geometry in their ability to generate a mental form. Pound talked about the progression from the arithmetical to the algebraic to the analytic geometry as a paradigm to understand the generation of form. The arithmetical and algebraic don't make a picture, but an equation for a circle governs the formation of a circle.

He finds in the idiom of analytic geometry where points of space become sequences of meaningful particles defining a form when they are yoked together in an algebraic equation, a paradigm, for the kind of poetry he wanted. He needed this idiom to create. Here words are the points being entrained by the syntactical pace. He said that the difference between art and analytical geometry is the difference of subject matter only. The subject matter of art is the complexity of human relationship. He talks about how this abstract approach:

---

[1] I worked with this material extensively in the book *Knight of 1000 eyes*.

(x,y) (x,y,z) (corners, boxes, dreams) (vertices, edges, faces) (mask, ego, super-ego) does not interfere with spontaneity and intuition, and that he saw where to put the line from intuition in this new idiom. My plan was to build a solo theatre piece. In the style of what they called Dream Theatre or what might be called Mundra. I hoped writing theatre would bring that sculptural roundness of voice. I figured the best way to learn something new is to start out on a project and ask for help along the way when you need it. It seemed simple enough: a solo psychodrama to see inside my own mind.

I had my own theories about theatre and wanted to test them in the experimental laboratory theatre. It was like a kind of neo-cubism. We could see these things in the visual arts, Kandinsky and Pollack, and now here they were being done in theatre in Artaud and Growtoski. The term Performance Art had not yet been invented. Theatre is a simulation of my reality, consecrated into the 4D time cube of the theatre. The tesseract. To build a real-life fully extant and reified tesseract. Where else but in the theatre? Made out of little tesserae, of space-time tiles of motion and emotion motating gesture.

A tesseract is the simplest closed four dimensional object. It is represented visually in the flat 2-space of the page by a cube within a cube. Imagine one corner of the inner cube. In this corner, the three dimensions of space: length, breadth and height come together. Coming out of this corner is another axis: time. Time is shown as a 4th dimension, perpendicular to the corner of this little inner cube. Then this 4th dimension shown perpendicular to the other three, on all the corners of the inner cube, is connected corner to corner to enclose the object in time, like time was an extension of space, lofting up, out of the inner cube. The abstraction is difficult to imagine in just words. Certain artists, in this case an actor, can give us a more lucid intuition into what the tesseract is. It is a model.

An example from T. S. Eliot, in *The Waste Land* shows language in the service of a hallucinatory image.

A woman drew her long black hair out tight
And fiddled whisper music on those strings
And bats with baby faces in the violet light
Whistled, and beat their wings
And crawled head downward down a blackened wall
And upside down in air were towers
Tolling reminiscent bells, that kept the hours
And voices singing out of empty cisterns and exhausted wells.

The syntax of this poetry is like a snake traversing a higher dimensional object.

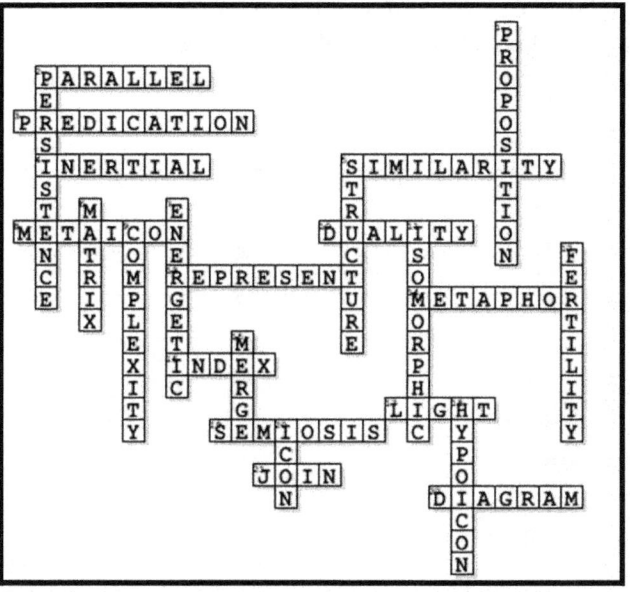

The higher dimensional object is a matrix in the abstract:

```
     A B C D E
A    0 0 0 1 0
B    0 0 0 1 0
C    0 1 0 0 0
D    0 0 0 0 1
E    0 0 1 0 0
```

1 if the edge is there
0 if no edge is there

This model sketches an intuition that is both scientific and mystical, because it comes as a gestalt shimmering in and out of the space time of your reality. You must get it in your own time. It will have as strong an influence on your world as your world will have in supporting it.

To help remember my lines, I thought to create a Mnemonic Theatre and then map the images onto the vertices of a hypercube – a four-dimensional cube. Humans cannot naturally perceive four dimensional objects. Our ancestral needs and evolutionary pressures limited us to three: length, width and depth. We can see three-dimensional cubes with ease, whereas a two-dimensional creature — a Flatlander — could not. But a 2-dimensional creature could see a 3-dimensional cube if we unfolded it and laid it flat. We recognize these images from standardized tests, like the SAT.

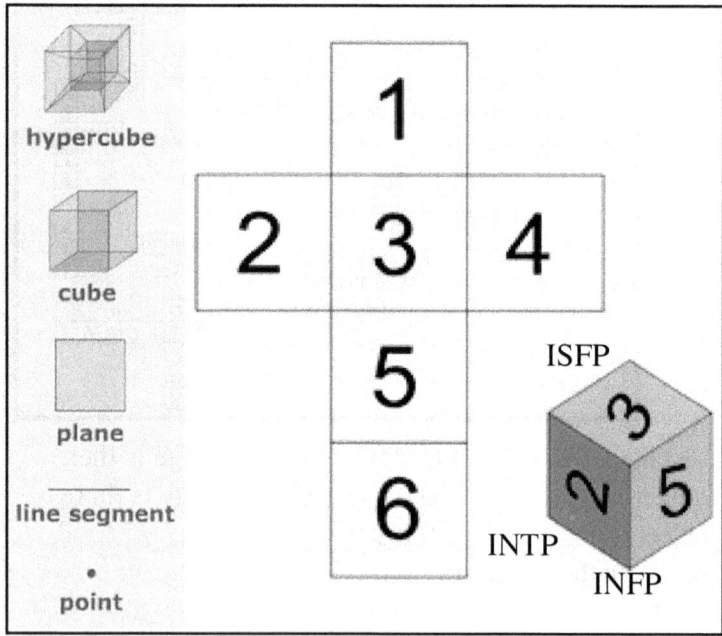

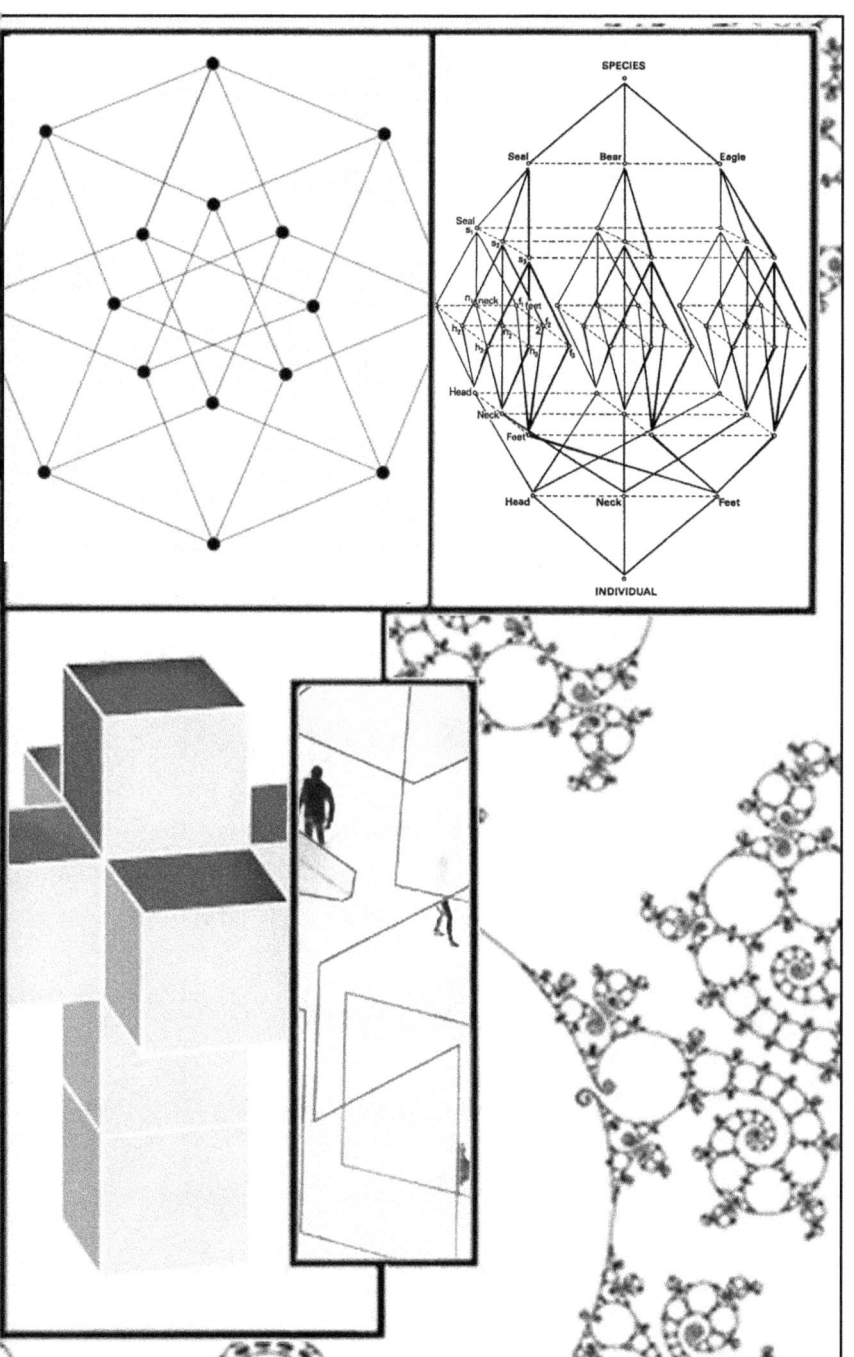
SPECIES
Seal
Bear
Eagle
Seal
Head
Neck
Feet
Head
Neck
Feet
INDIVIDUAL

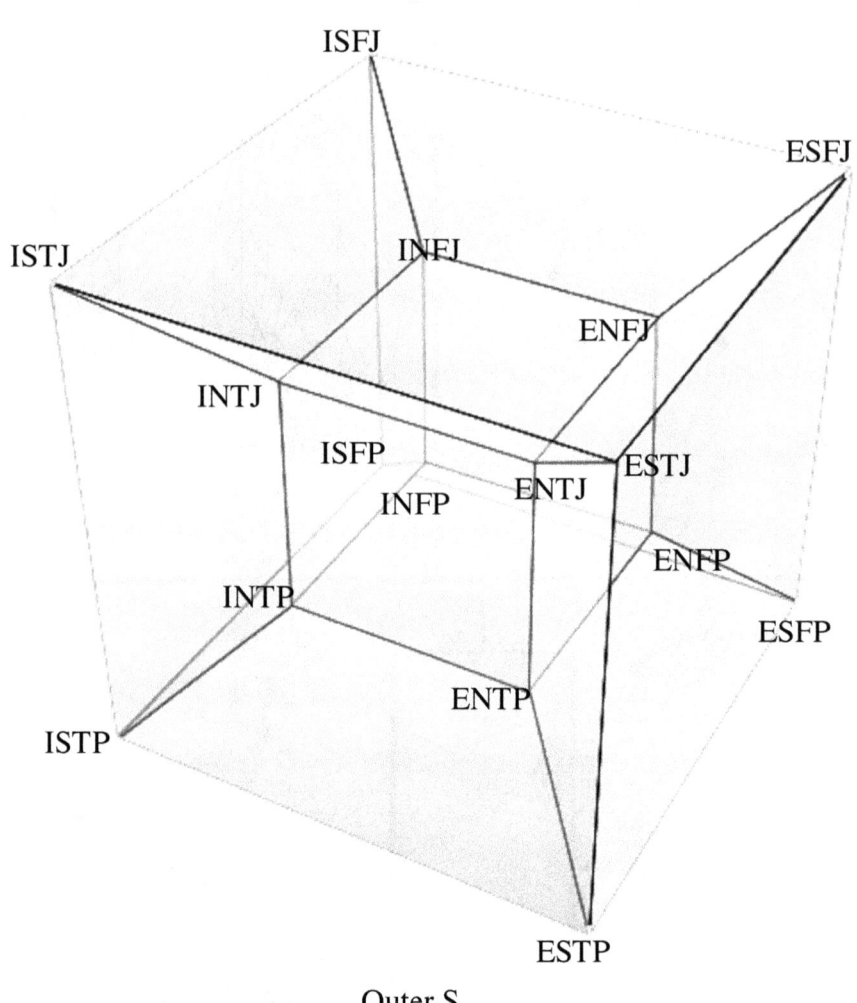

Outer S
Inner N
Front (Right) E
Back (Left) I
Top J
Bottom P
Inner Top T
Inner Bot F

4D Model of Intrapsychic Personality Manifestations

The model at the left uses the (dynamic) tesseract as a constellation of dimensions upon which the Meyers-Briggs Type Indicator (MBTI) is mapped. One of the 4 dimensions of personality type (I—E, S—N, T—F, J—P) is assigned to a dimension of space / time, which in the abstract is the tesseract.

A dimension is a continuous variation between opposite polarities of consideration. Between Judging to Perceiving, between Extroversion to Introversion, between Sensing to iNtuiting, between Thinking to Feeling. The player can control his movement in a dimension.

In this model the Outer Surface represents sensing, while the Inner surface represents Intuition (N). The front face to the right represent pure Extroversion, and as you move further back to the left you get closer to pure Introversion.

For example, Hemmingway was more Introverted than Jimmy Stewart. Harry Truman more Sensing than Marilyn Monroe who was iNtuitive.

The outer Top is Judging and the outer bottom represents Perceiving.

The inner front represents Thinking and the Inner back represents Feeling. These attributes form the personality, which is here visualized as a join of the possible spaces. A person can inhabit the spaces of personality.

The outer forth dimension and the inner dimension represent the Sensing Intuiting axis. These 8 independent cuboids can rotate, and reflect and translate bringing different personality types into contact with their opposites.

Pages 234, 261, 277 etc. show different transformations of a tesseract. Mathematicians have done quite a lot of work in this area and the psychologist and semiotician would be well served by availing themselves to the implications of these models.

Just as you can follow the traces of how knowledge gets into the personality, you can run it backwards. The artist can inhabit different coalitions of sub-personalities to further your agenda on the outside world. Or in pursuit of a meditative state, feel these filters of mind letting go as you focus on them and then remove focus. The way in and the way out can be accomplished in the 4 directions and hinges on a few key questions. Here starting with energy source.

#1, Are you energized by the external or internal?

#2, Do you perceive through sensing the external or intuiting the internal?

#3, Do your prefer Thinking or Feeling in making your judgements about what you have taken in?

#4, Do you prefer the immediate gratification of quick judgement or are you OK with waiting for the gifts of perception.

As you can see, this isn't just a checklist of the attributes of a personality's processes. Each answer hinges on the one prior to it. The possibilities unfold like a space-time origami crane. How could each of the distinct viewpoints interact with each other and come to a consensus? It's time for a chart here. Ladies and Gentlemen, loyal inquirers of the Church of the Coincidental Metaphor, I present to you:

The Four Dimensional Matrix of Personality Types

		4	Sensing		Intuiting	
1		3 2	Thinking	Feeling	Feeling	Thinking
Introvert		Judging	ISTJ	ISFJ	INFJ	INTJ
		Perceiving	ISTP	ISFP	INFP	INTP
Extravert		Perceiving	ESTP	ESFP	ENFP	ENTP
		Judging	ESTJ	ESFJ	ENFJ	ENTJ

Here's how to use this Four Dimensional Matrix of Personality Types. Simply complete the following sentence:

I get energized as [1], taking the world in by [2] and [3] about it and that leads me to be robustly [4] in time.

[1] = "Introvert" or "Extrovert"
[2] = "Sensing " or "Intuiting"
[3] = "Thinking" or "Feeling."
[4] = "Judging" or "Perceiving"

For instance, Ghandi was an INFJ. And his operational mantra would read, I get energized as Introvert, taking the world in by intuition and having feelings about it and that leads me to be robustly judging in time. Walt Disney ESTJ would approach the world by this principle: I get energized as an extroverted working with people, taking the world in through sensations of movement and sound, and following the story of thinking about it and that leads me to be robustly judging of the progress of the work in time.

This is just one use of the 4x4 structure[1].

Overview of this book's Semiotic Reflections

A similar program using the Tesseract model, can be undertaken to help follow the levels of Sign and counter-Sign (generative, negating questions) passing throughout this text. This is accomplished by considering a plane of the tesseract to be a Semiotic Square. (Jung's Personality Types, can be seen as based on the semiotic square.) Thus semiosis is modeled by considering the tesseract to be an ensemble of planes upon which are projected semiotic squares. (See diagram page 283.)

In *The Subliminal Kid* we are presented a play which is a miniature of the novel, written by its main character, Walker. The novel plot shows the evolution of the main character Walker as he starts to loosen up in the company of good friends, drugs, love. These forces are embodied by Wild Bill

---
[1] See *Knight of 1000 eyes* for modeling the ancient I Ching.

and peyote and Laura. The novel is a deepening of a modern education. The chapters are a series of lessons sorted into two groups and we can see them as symbolic answers to two kinds of questions.

The main question driving the semiosis of this novel is: What if you for once in your life let go and just feel all you need to know. In the novel Walker negates the admonishment of being busted by the state and continues to grow pot. He creates a post nuclear scenario that negates the nuclear dread of a generation brought on by the cold war detente of MAD by gaming post holocaust survival. He negates a grandfatherly admonishment meme / curse. In the novel Walker is moving toward a more primitive natural way of being. In the monologue the character is moving toward an advanced computer mind interface. The play has the character exploring the negative of his slacker ways looking for a job. The novel has the character exploring the slackening of his logo centric ways.

The most interesting phenomenon that the reader experiences is being present in the moment of creation as we watch the writer struggle to create. These chapters where Walker is trying to compose a love letter, where he wants to celebrate the 10 year anniversary of the moon landing, others. We start to have the feeling of regarding past events through several sets of eyes: we have the Walker in the moment of the story, the earlier Walker of the memory in the past, an Uber-Walker writing the story, the story of the aesthetic of the play and the play. Look for riffs around who is looking through my eyes. There are different time lines, too: the one of his education, the one of his progression into the indigenous spiritual world view, and the one of his maturity contrasted with his young girlfriend. The aesthetic of Shift—, Switch— and Stream— move around the semiotic square to manifest the illusive 4th term: the Self.

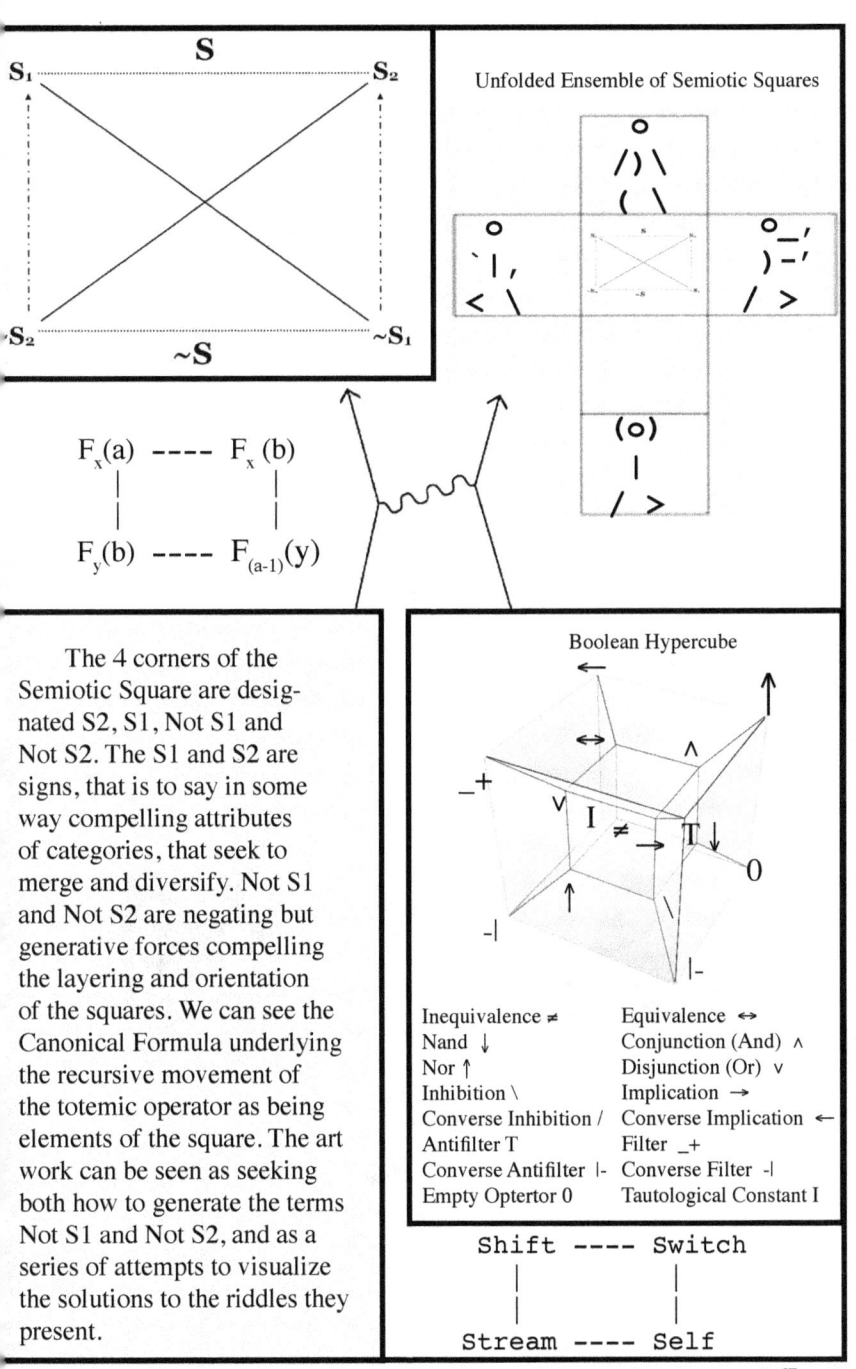

The 4 corners of the Semiotic Square are designated S2, S1, Not S1 and Not S2. The S1 and S2 are signs, that is to say in some way compelling attributes of categories, that seek to merge and diversify. Not S1 and Not S2 are negating but generative forces compelling the layering and orientation of the squares. We can see the Canonical Formula underlying the recursive movement of the totemic operator as being elements of the square. The art work can be seen as seeking both how to generate the terms Not S1 and Not S2, and as a series of attempts to visualize the solutions to the riddles they present.

Inequivalence ≠   Equivalence ↔
Nand ↓   Conjunction (And) ∧
Nor ↑   Disjunction (Or) ∨
Inhibition \   Implication →
Converse Inhibition /   Converse Implication ←
Antifilter ⊤   Filter _+
Converse Antifilter |-   Converse Filter -|
Empty Optertor 0   Tautological Constant I

he Varieties of Logical Experience, or Semiosis as Primary Algebra, or $IF^{IF} = IF$

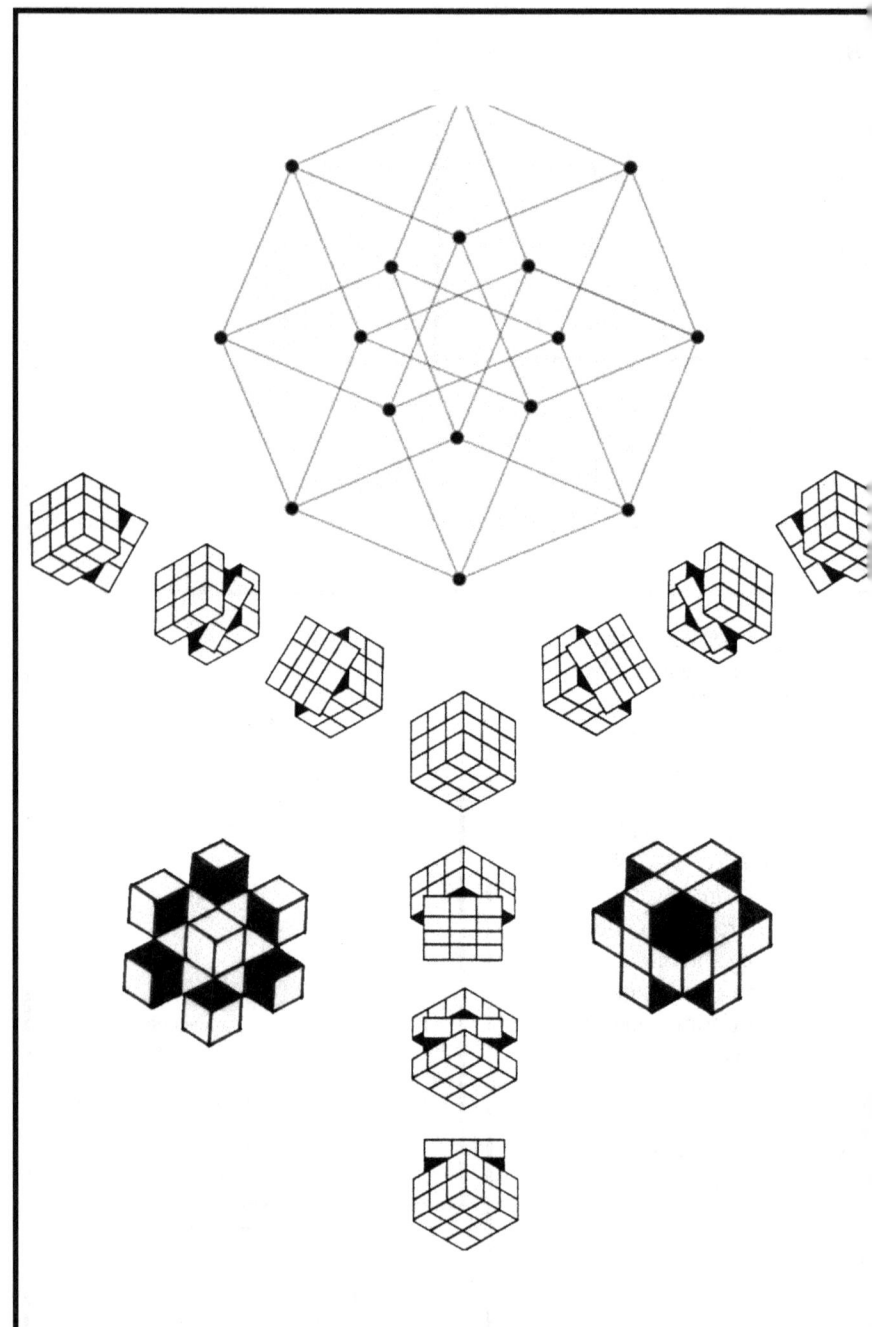

The Monstrance of Forking Paths in a Separable Space

Quarking the Cube

The cube floats in the space of my imagination now, gracefully drifting and tilting at odd angles, out there in the periphery of the past. For a while I was obsessed with it. It came into my life at a time when I really needed a gripping avocation to take my mind off a sense of futility and defeat that was coming to haunt me, after turning 30 and not getting anywhere.

For now just picture it as a Rubik Cube with 6 faces that are colored WBGYOR: White, Blue, Green, Yellow, Orange, Red. Looking down on a corner of it from above you can see three sides. The other three sides are hidden. If you look at a face, you can tell some of the other colors of the four faces touching your face.

The Cube is composed of 6 faces, each face made up of 9 tiles or sub-cubes in a 3x3 array. The faces of the cube are connected through their centers by an ingenious universal joint which allows movement in the three directions of space — vertical, horizontal and through. Mathematicians have computed the number of possible arrangements of the cube (permutation states) to be very large, though still finite. One conjecture has a blind Cubist performing rotations, twists and slices at one per second taking some 3.78×10^{15} seconds to go through all the different possible cube states. This is 120 million years. More than enough time for the dinosaurs to evolve, go extinct and for the little shrew-like mammals around at the time and newly liberated from under the terror, to evolve into man. So when you took hold of the Cube, you were indeed holding infinity in the palm of your hand.

Rubik the designer was an architect and he developed the toy to help his students think more concretely about building. It certainly came into my life at an opportune time, for things were getting mired in abstraction.

When it came to the cube, I had not put my childish ways behind me. I spent quite a lot of time in day dreams about the cube as some cool 4D architecture building I would some day be able to inhabit, perhaps situated by the sea or on a great plain or steppe somewhere with a commanding view of vast distance. Or just thinking about a color in the scene out the window, and how through something behind the surface it came to be here. And that machinery behind the surface perhaps relates to something of another color, or attribute of something else.

Each face of the cube has 9 little tile cubies. We called them **cubons**, a name Lyle Burnett and I came up with, in honor of particle physics — photons, electrons, muons, protons populating the subatomic zoo, working their way up the ladder of being to the all-important, multivalent, Protean, — carbon. The center cubon designates the color of the face. The made state of pristine symmetry when each cubon is in the correct position and orientation on its face, is called Identity. The (group) operators take you away from and toward the Identity.

Each 3-by-3 matrix (or "slab") of these nine little tile cubes, or cubons can be independently rotated (or "twisted") on an axis perpendicular to the slab and passing through the center of the slab. + The two slabs going through the center, we called the "slice." And they were on a separate axis from the face, perpendicular, to the face. (They slice through the center of the faces, creating an axis of symmetry). Thus it is natural to follow the symmetry and use the fact that the cube is a separable space.

It was Lyle who introduced me to the cube. He brought one to the school where we worked, Austin Community College. I worked as a physics lab technician setting up experiments and fixing damaged electronic equipment, and he was an Physics TA running the program and tutoring.

Soon I got one; they came in a little plastic cylinder with a tiny little fold-out page on what to do. We'd hang out at his place, a duplex on Depew Avenue. Sometimes we'd smoke weed and work on the cube. His was a typical Austin pad: book cases made of cinder blocks and boards; lazy boy recliners that were so bad cut up you had to cover them with cushions; a formica table in the dining room leading to bare little kitchen with an empty fridge; interior in style of hippie freebox. We didn't have a car so we took the free shuttle bus or walked, wearing bill caps to keep the relentless sun out of our eyes. I did so much walking in those days. I would try to stay in the shade. I called myself a shadow-seeking heliophobe.

LB — I began calling Lyle Burnett that after we started hanging out, 'cause that's what they called him as a kid back home in the small Texas town he was from. (I joked, LB was short for pound because we were smoking lots of weed. Then: "He ain't heavy, he's my brother.")

LB and I worked closely on the cube. We were like mountain climbers, "tethered together" as we ascended Mount Bourbaki. We had actually been wandering around in the foothills of Mount Bourbaki since the 2nd or 3rd grade for we were the first generation to get started in the New Math. He was great with hands on, loved to just experiment, and I kept getting these intimations of group theory and abstract algebras and analogies with fundamental processes. We visited each other's pad often.

It was amazing that the Cube fell into our world then, for Physics was on a heady push to understand the deep structure of the universe by looking for the laws of nature in the symmetries of space and time and other dimensions. It was all so abstract. The old school had started with probability waves that at least gave an intuition of fields sending signals but the wave collapsed when you tried to measure them. Then

the new school had an observable as the result of an operator — a matrix — that produced complex numbers as dot product projections and the outcome of this measurement of an observable must give out one of the operator's eigenvalues. When you observed, that is to say reflected a light off of the observable, the photons in the light disturbed the state of what you are trying to measure, so uncertainty was fundamental. The probability wave collapsed and you see a spectral line eigenvalue indication of a transition in energy level. (Contemplating probability waves was quasi-mystical.)

Foresightfully, some experiments were done or the whole thing would have been a philosophical fiction. We actually calculated the speed of light from a split beam laser interferometer and a fast tracking oscilloscope. And my favorite experiment would have to be the Millikan Oil Drop experiment. Looking through the spy glass at this lovely bubble, falling in a fluid. It was like looking at a distant planet in the firmament. The tiny spherical bubble had light and dark sides. You could almost see little oceans on it reflecting the moon light. We used Reynolds number to calculate its diameter as it moved through a non-turbulent viscous flow. Then applied a field to stop it in its tracks, make it move up and down. Holy Gort. Klatu barada nikto! Don't make the world stand still, the shear forces alone — from the deceleration, would make everything fall down. Yeah, that one tickled me.

Coincidentally, since I had come into Physics from EE, I had some background in circuits and the idea of incoming scattering around this lumped element network (or tree, or group graph) and phase shifting the signal was not new. And I had a lot of experience with Boolean Algebra — used to get the state tables of switching circuits, logic gates, oscillating flip-flops. This elegant and friendly representation was good old symbolic logic as variables in an arithmetic over a field of binary numbers. And it was another paradigm shift moment

to see how the computer was like our binary mind at its most basic core.

I bothered my mind quite a lot with all these concepts and metaphors from physics. They are some of the great ideas of my time and the artist needs to play in this field to appreciate more deeply what is the nature of nature. School presented it as axiomatic, a finished structure, you did not get the context unless you dug down into it. In academia it was all such an abstract different world view, and you better keep up with it or you were out. And there were not that many teachers who could or who would, take the time to help you settle in to the zeitgeist. And yet it is just as simple as a coke machine or as real as a toothpick dispenser. Suppose you are baking a cake and you need a toothpick to poke into the center to make sure it is cooked. You have a toothpick dispenser and when you shake one toothpick out from the whole ensemble of toothpicks, you expect 1 or sometimes 2 to work their way through the tiny holes in the cap. The toothpick dispenser is like an experimental apparatus and the toothpicks inside are all potentially able to come out the hole, shaking and pouring is an operation that projects some situation into existence out of these probabilities. Every moment, everywhere, the sorting demon is haunting this world with its endless making of (irreversible) distinctions. The universe is a machine. Press a button, signals get sent, something happens. And we were in it too. It is as simple as hypothesis < > evidence or as they say in quantum mechanics, <hl :<hle> : : lo> : <olo>.

LB had been an an x-ray crystallographer as a graduate student, and he had worked in oil well geology, so he was a really good down to earth experimental guy. I had taken classes in vector spaces and algebraic structures and the cube was a godsend to get your hands on a physical representation of that stuff. The Cube was a gedenkin laboratory to study

the kinds of thinking involved in math problems beyond number, and getting the answer.

This Cube representation was timely as I had studied *The Vector Space Theory of Matter* from the guy who wrote the book, F. A Matsen. He would wax maniacal, excited about the *ab initio* calculation, starting out from just pure group symmetry considerations of space. Matter and antimatter emerged from the *ex nihilo* of the vacuum to be governed by the laws of symmetry — to represent the laws of symmetry — as symmetry held sway over domains until the domain boundary was reached then symmetry broke and became another symmetry. These theoretical physics types are idealist — Platonic Pythagoreans who want to be able to derive the behavior of the world out of the symmetries of space and time. The laws of physics are such symmetries: Conservation Laws. Conservation of Energy is due to Time Symmetry. All energy must be accounted for from the beginning to the end of the change in time. Conservation of Momentum is due to Space Symmetry. All momentum including rotational, must be accounted for at the beginning and at the end of a motion, (translation, rotation, spin) in space. All momentum must be accounted for at the beginning and at the end of an orbit in space. They could design vector spaces that had the necessary properties to map onto the observables. These are the energies and moments that show up. The operators represent experiments acting on the elements of the space, the vectors. The vectors carry potential information that is projected out in the form of eigenvalues — characteristic energy signatures (spectra). If the experiment is set up propitiously, that is if the operators are symmetry adapted to the natural way the vectors or states of the system are organized, then they will produce these signatory spectra which tells you that you have a judicious insight into the elements of the space.

They reasoned by Classical Analog. That is to say, the well

known functions of Newtonian mechanics were now thought to be special cases of operators. There are still orbits. And orientated rotations. In the quantum world when orbits are traversed and orientations flipped, discrete energies are given off. And these eigenvalues, colors, are the measurable traces of structural change.

These symmetry adapted operators are global requirements empowering (or getting representation) in local phenomena. Myth is also such a global operator that locally, in an individual, becomes archetypal transference.

The preceding paragraphs in an example of basis change. Or an analogy. I got going on this idea of a basis change. Physics was an art of sketching in an energy landscape, like the beautiful vector diagrams of Kandinsky and Klee. We'll get back to that later.[1]

But at first the Cube was overwhelming, intimidating. The feeling of sheer free-fall fear of not having some ready-made, or at least potentially understood algorithm solution was intense. It exposed your ego to emanations from the core. But LB was the go-to guy when I got stuck. He'd take the cube from me, squint hard at it for a minute, say something like, "Oh, you need to apply the triple edge swapper here — on these three pieces." Then give it back to me and watch to make sure I did the move to make the change in the Cube's state. I gratefully accepted. When I got stuck by myself, I usually had to take apart the cube, down to the universal joint

---

[1] The idea of characteristic value, (or eigenvalue) is the basis of metaphor. For example a stanza in Shakespeare, Sonnet LXXIII:

That time of year thou mayst in me behold / When yellow leaves, or none, or few, do hang / Upon those boughs which shake against the cold, /Bare ruined choirs, where late the sweet birds sang.

The LIFE IS YEAR metaphor explores the analogical similarities between the characteristic values of a year and a person's life. A year begins and ends, has seasons. Thus we are able through metaphor to traverse the dual space implications of the seasons of the year and observe them in a life. In <o|**H**|o> one is seeing an isomorphism across ontological categories.

going through the core and reassemble it in its made state.

Really, I felt like I was LB's student; I got into the wake of this interested guy whose mind could range into anything and followed him on this curious quest into the cube. But LB remembers things different, god bless him. He says that when he first handed me the cube, I quickly discovered the Slice, where you twist the right face AND the center and bring back the right face. This leaves the center sliced through the 4 cube faces it touches. LB said he had never seen it done that way until I casually manipulated the cube that day. He put his hand out and stopped me and said, "What did you just do? Can you do it again?" Fortunately I could.

This got him moved away from a cube coordinate system based on the fixed position of the face-center cubies. The slice notation is much more symmetrical and compact. It was easier to visualize and remember. It's fairly standard usage now.

The cube itself is a separable space, like how von Neumann in his axiomatic treatment of Qantum Mechanics showed. In our cube representation of this, operators involving the slice did not effect operators involving the corners but did involve the edges. So we looked at it as three groups: the corners X the edges X the slice. The interaction of symmetries is like the situation in the Standard Model where $SU(3)$ x $SU(2)$ x $U(1)$ spontaneous or directed symmetry breaking have a configuration that distributes itself into symmetry domains described by these group and their algebra of operators. These operators are the spontaneous or directed (experimental) activities of taking measurement.

I always felt I was way over my head with these natural born science types. I realize now they can be as passionate as any poet and even more easily misunderstood. I came into physics from literature, I was not naturally one of those types of guys, who were always taking things apart or figuring out codes from alien or other universal communicators. I had

studied groups to help get through QM, and I was delighted to see the toy Cube as a representation of a finite group algebra. I studied math like you would a foreign language, to be able to read the literature as a form of rhetoric.

But as I mentioned, I was slipping into some dark times back there in Austin. No love in my life, no prospects. LB was a guardian and an inspiration at the time. LB inspired me to dwell in the Cube. The job at Austin Community College was the first job I ever held for more than a year.

I was casting about, had gone into a sexual dry spell, an off month that dragged on and on for several months then was stretching into a year. And there was no immediate sign of any change going to be happening any time soon.

I was still horny, still enjoyed fantasizing about females, because it is the otherness, but after awhile pure otherness gets to be a drag. I wanted someone who could relate to who I was and, who I saw myself becoming. I was a mess, kind depressed, OK? Slipping into a dark hole, looking for a way out. Time to get out of Texas.

My general malaise and cognitive discord might have come from entering the shadow world of mycological experiments. At the Parkway house I was raising psilocybin mushrooms. Pressure cooker, scrubbed kitchen. Inoculating the substrate with mycelium on a flame-sanitized loop. Getting all scrubbed, putting on a hair net. Wearing a surgeons mask to cover my beard. Gathering spores from the basidocarp. I was slipping into nacreous fourth kingdom darkness in those days. I bought the mycelium through mail order from a doctor in San Antonio who advertised it in *High Times* magazine.

LB and I would partake of the Divine Soma over at his duplex on Depew. Then, possessing psychomimetic kinetic energy, we would go on long treks across campus, past the Drag, down Shoal Creek, through Clarksville, to a midnight

swim in Deep Eddy pool and back.

I had never really understood it before but I was involved in a different script, or mythology; we called it tapes back then. I suppose that by Texas standards we were perhaps somewhat less masculine than the standard issue. We were "nontraditional" men. Basically "nice guys". We thought of ourselves as "new men". We were pro-feminist. I had learned a thing or two from the women I had lived with, especially getting over all the jealous possessiveness — that was kind of weird. This human potential movement mindset starts all the way back in the anti-nuke and the anti-war movement. Certainly The Draft was a black hole. Following the traditional macho male hero script into it, could take you to your death, on a foreign soil.

Jung helped us to understand ourselves. With his understanding of the animus and anima parts of the psyche. He certainly is a prime mover of the human potential movement. We were trying to find our way into reclaiming our true Self from the very narrow set of behaviors open to us. It was alright to try to be in touch with your feminine side. It was being integrated. "Caring" and "emotionality" were not to be just left up to the feminine psyche of women any more. How on earth did all that get started anyway, Male and Female tapes. Recorded messages that you played over and over again in your mind. Television and video were the agents of myth; they were good metaphors for the meme. Though I had grown up with the 4F Club mentality — find 'em, feel 'em, fuck 'em, forget 'em — I, like most normal guys, wanted a deeper more caring intimacy than that. To attain that state required vulnerability, alas.

We were definitely not homosexual, nor feminine — that is to say not into being compliant, dependent, or submissive. We were not into having relationships with men that are sexual or overly intimate. But it was good to be able to talk

about anything with a friend, and even appear vulnerable and lonely. We were beatniks, bohemians. We were hippies, with hang-back slacker tendencies. Let the alpha males rush on in there, they can have the glory, whatever that is. We were not dandies, or fops though I did have long hair. I had it for being confrontational, a statement that said I was not corporate or military. We did not devote extra time beyond basic hygiene to looking good. The grinding away of financial power left me looking for other models of maleness among blacks and latinos rather than whites.

It was OK to be teaching young people about the beauty of physics, and I really liked hanging around with the very fun and intelligent people in the Physics Dept. But then one day I was teaching this arrogant whiny Iranian guy, who obviously thought he was superior. (ACC took in students who were not ready for UT, and tutored them and worked with them generously.) But this day the student was showing me his watch and when I realized his watch cost more than I made all year, the caché of being a teacher — rising to the challenge of having to stand and deliver — suddenly lost its luster. I mean it just vanished like it was vacuumed out of me. I literally felt my spirit collapse in that sucking sound you hear when the insight of the exploited suddenly sinks in.

The Cube began to take on mythological powers in my mind. When you held it up, in front of the sun, it was about the size of the sun, and it was a representation of the energy fecundity that our Old Sol was, being partitioned into many realms, for the sun floods all niches with its free energy.

Somewhere out there in the infinite void, intelligence was falling through this little hand-held 3D axis like it was a quincunx funnel of new understanding and growth. Enhanced rhythm, physical eye-brain coordination, a sense of place in the economy and the universe.

The music composers really understand. Working with the scrambled Rubik cube is like composing, or creating. It is the natural way the brain, a neural net, likes to interface cue and intuition. For example if you focus in on a particular color or corner, you will see that it might relate to some other cues around it. This is abductive reasoning, a method of logical inference introduced by Charles Sanders Peirce which comes prior to induction and deduction. It is a "hunch". You could feel the ludic play of mind in the entrainment of hunch. One starts to read signs of the emergence-into-existence of a pattern.

The Cube is holistic. In its disarray, it is a field of potential cues. You go into a kind of concept field without knowing in the beginning, how many pieces there are that will have to be moved. You have to "see" so that they match each other. Then one logical decision step follows another or is followed by the next. It is the start of a group. Out of this group of field elements, a cue occurs to act as a conceptual attractor, swirling up both the hints and a possible strategic direction simultaneously, for the Cubist to take. These are partial perceptual gestalts of which concept-pieces (cubons in relation) are necessary for the more final picture — for the whole concept system — to fall into a proper alignment or beautiful relationship. These hints are elements of the symmetry group and will represent which ones should be rejected, or are subsidiary or are even elements of other subgroup symmetry. The desired gestalt gradually emerges during the process of finding and matching the cues.

The Cube can represent any field, or domain or system. Cues emerge from a field. This field in general may be any external automata system capable of containing suggestive patterns. Here are a few suggestive fields: concepts, information, knowledge, intuition, science, paradigm-shift, science-fiction, myth, folk-lore, lies, truths, ideologies of any sort, and whatever helps you organize your science and or fantasies. And

of course, stimulates. The cue's function is that of (concept) attractor that invites other covariant concept trajectories from various fields to form a coherent whole.

To sum up then in general, the work of a Cubist is to find the right concept-pieces and match them together to form an energy structure. You wade into it not knowing how many pieces there are, or how they match each other. They might be a group, which is a very minimal structure on the set. Some elements in the group — the group generator(s) are by far the most important. (In QM the most famous one is the anti-commutator from Heisenberg [P,Q] — the Uncertainty.) In the Cube it was the [RU] commutator. In the cube they act as conceptual attractors giving the Cubist hints of which concept-pieces (group elements, field numbers) are necessary for the concept system to manifest its Identity in a state of harmony.

Sorry to be so abstract. We were fervent and furtive aficionados of the Cube. After a while we started getting good at it. I could routinely do the cube in under 2 minutes which is not bad for an adult.

We had a little dance that we did, crazy guys. We called it quarking. Quarking the cube. Sometimes we'd just get all crazy and joyous and start doing the dance while manipulating the cube. High on weed and quantum mechanics, we felt like we were attendants of the new probabilistic inference engines, emerging into the zeitgeist. The quarking dancer might start by doing a twist on the cube with his hands and then rotates his hands and undoes the twist, just to start the motion rolling. He does a slice then undoes it, or does a long sequence to bring the cube into a pattern, like Dots or Snakes. The cubist is doing a performance. Singing a little song about, "Come on people we are quarking the cube."

"Rotate around the central axis. Slice through the center. Slice to the right 4 times. Real smooth."

And slide to the right 4 times.

Or when one was finishing, resolving the cube, he would do it with a flourish, shaking his shoulders and present it out to you: "Done!" And say emphatically: "Resolved!"

Or "Returned to its state of pristine harmony!"

Or "Renegade potential in détente!"

It was delightful. One of us would jump up and do it for a while then sit down and the other one would jump into the center of the living room. We are wearing skinny old blue jeans and it was so hot we usually wore only a T-shirt or a red checkered shirt. Of course we did not have air-conditioning. Austin slackers.

We were doing a lot of Tai Chi at the time. I had been teaching LB some of the first 19 moves of the Yang Style Tai Chi long form that I had learned in Berkeley. I organized the class because I wanted to get to use the beautiful little ballet studio upstairs at the Austin Recreation Center on Shoal Creek adjacent to ACC. And I wanted to explore movement. This studio was a small ballroom with a bank of windows all along the north wall, and hardwood floors. It was lined with mirrors on the other walls, in front of which ran attached parallel barres for stretching. In order to get ownage of this room I had to offer a class in Tai Chi for the public. I made a strange looking sign for the class, it had a finger-painted ideogram / mandala of some multi-footed icon-man skittering and skiffeling in a fluid environment. We didn't get too many students. For a while there, this beautiful room became like my studio. I was living just across Lamar Blvd. I had a tiny attic apartment on Parkway, it had "waterfront" on Shoal Creek. In fact the bottom floor of that house got flooded that year when the creek rose way above its banks. They have a high water mark on the Whole Foods building on Lamar. That was also the year John Lennon was taken from us.

LB and I got into doing this quirky Quarking the Cube

dance in which there are isolations or jerky frames of mime motion from one still vogue pose to the next. All the while holding onto the precious cube and showing it like it was a monstrance. This reflects the idea in Tai Chi of keeping the hands rounded in holding on to the ball, an imaginary circular enatiomorph (yin/yan vessel) at the *tan tien*, or centroid of the body.

It was mirthful doing these exaggerated staccato mime poses. Then a quarker might get distracted and stand there doing the cube, saying out loud the operations: "RFR'F' UR."

And pronouncing their names: "Corner swapper; double edge flipper. UFRU^{2}R'F." It was as though intoning the concatenations of the operators of this algebra, brought order into the divine emporium — as it was emerging from the densely packed realm, of convoluted potential instantiations.

The dancer speaks: "The human body as Cube. Arms and shoulders left and right slice turn left and right."

Below the waist is the Down. Above the waist is the torso, twisting and turning, and the head is like a corner cube. Right!

"To the Right." He'd turn to the right, then, "And up." Learn over and rotate his torso.

"To the Left." The player showed the hands moving that face. Calling out the moves. Taking a step to the left.

"Slice up; up; slice up; up-squared; slice down; up; slice down; up-squared; slice down; up-squared.

I put on a pair of sunglasses: "These are some kind of special glasses to allow one to see the ancient Parmenidean cube at the center of the world. It was the only one."

We needed a name for ourselves and what we were looking into. We thought of Radio Ranch Underground Empire of the Air like out of the old black and white Gene Autry serials we had grown up with of Saturday mornings. Another suggestion was New Science Laboratory. But the one I had to have was Gruppenpest. Yes! We were gruppen pests quarking the cube.

As I explained to Lyle: "These were the people — Weyl, Wigner, von Neuman and others who were axiomatizing quantum mechanics with concepts from group theory. The previous generation of physicists who had grown up with the concept of the continuous wave equation, were now being eclipsed by the insights of this new discrete algebra that was very abstract, it seemed to be just an ad hoc algorhythm for relating data to theory. The old guard called these new followers of Heizenberg using formal operators on a vector space: "gruppenpests" as a derogatory term. Later somebody showed the matrix and the wave were equivalent."

In the voice of the southern new-age preacher, I announced my mission: "I am trying to convert you to the 6-Fold Way." I made the sign of the cross and announced the 6 commandments of this new quantum religion: "Model the Hamiltonian, symmetry adapt the vector space, map the data to the Hamiltonian (do the experiment), compute projected eigenvalues, see the symmetry delineated eigenvectors, and arrange the characteristic observations into signal identifiers (quantum numbers.)"

Yes! We were gruppen pests quarking the cube!

LB had a couple of Lazy-boy recliners in his place and we'd talk about all kinds of stuff or just be silent, so all you heard was the little plastic cubons — click and glide past each other with minimal friction, for we have lubricated their surfaces with a dab of Vaseline. They click into place with a little shudder. We had sanded down the little ridges left over from the manufacturing process on these inner surfaces. The cube action must be smooth, maxed-out for speed, full race.

The movements of the parts of the Cube can pop out, pop in, orbit, spin, flip its spin.

Of course for Quarking the Cube, we picked up some licks from the various street dancing seen on TV and the odd player down town or at City College. (We walked all over

Austin. Eating mushrooms gave one tremendous energy, you couldn't sit still.) We could do the freeze and isolations like they did, but we couldn't put BOTH hands down and go to the ground like them because we were carrying the cube. But one did manage to put one hand down while holding the cube up with the other hand, and flailing, dancing like a Cossack dancer while holding the cube up like a trophy or something you were trying to keep above water. It was hilarious, crazy. We'd do these elaborate transition — some of these drops even evolved names: front swipes, back swipes, dips and corkscrews. The smoother the drop, the better. It was like the Cube moves. Or like the Tai Chi moves: Snake Creeps Down. Or, Step Forward, Lowly Genuflect and Punch Up into Opponent's Groin.

The Cube moves and having writ, moves on. We are Cube.

LB and I would carry on this high-speed conversation that included using scientific processes for the aesthetics of the world. They let one penetrate the moment and feel it more. For me, these concepts, metaphors were objective correlatives for being in feelings, for I do not distinguish very clearly between thinking and feeling. Sometimes I can't tell them apart. And physics concepts took me into a kind of semantic high. One entered into a conceptual system or a mythological space in which the entities were forces and processes. For me, sometimes my feelings were synesthesia, of feeling the universe. I was childlike that way. The sci led into the fi, led into feelings. I am not describing it very well here am I? Let me put it this way. At the intersection of psychology and physics, the body is afloat in a transcendental background, an "informational cosmos" which is the (probabilistic) basis of both sci and fi $\{<\psi|, |\phi>\}$. This background medium is the subject of both theological and mythological ways of cognition, in which the physical, mental, and spiritual layers, like the layers of machine, compiled and user language, create a

unity on the basis of a gestalting transcendental information / control language. This background has been staked out, surveyed and partitioned out by academic interests with special languages, Mathematics, Chemistry, Rhetoric and Cognitive Ecology and others. It is hard to say what's what in this environment anymore, especially with the establishment of Uncertainty as a cornerstone of fundamental science. I wanted to be able to walk into these intersections, for these are the places of epiphany and enlightenment.

When I say the Cube was mythological I mean it was a kind of Totemic Operator, or Embedded Analog Projection Coordinator — like the I Ching, or the Sepheroth or the Mayan Calendar or the Astrological Zodiac or the Buddhist mandala — a structure both internal and external upon which people could project their inner psychic processes and questions. It was an external memory for them, a mnemonic mandala into which could be stored part of an individual's mind like a peripheral memory, and that was useful to think with, like an external analog computer (or the web of Raven and Coyote.) One that you didn't think about, and that if you didn't think about held a whole lot of unconscious sway in the shaping of your destiny.

One day we were talking about the changes that the brain undergoes. "This is your brain. This is your brain on Cube."

What happens when you start getting assimilated into the Cube? Part of getting good at the cube is that multisensory integration is enhanced to support more intensive scanning of visual, spatial, configurations and integrating that into tactile stimulations. This is necessary to succeed at the Cube. Not that we would have been able to articulate it at the time, though we tried in our way, but success at Quarking the Cube called upon a speeded-up or improved semantic reception and concatenation into patterns, of visual, spatial, and tactile inputs.

If you keep practicing and playing as we did, the brain changes — like it does for any intense game player. The player scans more, and more intensely because one is trying to beat his previous time. The Cubist brain gets modified from the more frequent workout. The mind's capacity for ludic behavior becomes more lucid.

LB was telling me: "There are characteristic brain wave frequencies to compare across types of games. They did an encephalograph study and found the left temporal lobe in the cortex gets more of a workout. This area of the brain integrates the visual, the auditory and the tactile, with the internal. That means multimodal perceptual analysis is amped up.

"This is your brain on Cube," he said, (goofing on the anti-drug commercial with the line: This is your brain on drugs.)

"What does it mean?"

"It means the player is linking intention and action to memory of previous responses and the rewards associated with them. The Cubist is actively using this semantic / algebraic information to quickly alter motor plans for the hands and wrists. There is more hand—eye coordination.

"The brain starts doing a kind of synesthesia / math dance. Rapidly switching, across left and right hemispheres — the left associated with the hard wired logic and semantic memory; the right associated with emotion and socially relevant episodic memories. They found that Cube play is processed as a general ongoing cognitive puzzle rather than a discrete episodic game memory."

"Wow." To me it was like the difference between pulp fiction and lyrical writing. One was an entertaining story within the bounds of formula; the other so stimulated the brain that it became more integrated in the world. If that isn't post modern I don't know what is.

We were hoping these brain changes from playing the Cube generalize to performance changes in other cognitive

domains — working memory, processing speed, spatial reasoning. "This is your brain, this is your brain on Cube," became a kind of rallying cry, when we did our dance.

When it came my turn to do the Quarking the Cube dance, I might portray that early difficult, struggling lost stage. In my dance I would be Man Struggling Against the Cube, holding on to this tiger by the tail, leading me around. I could show being stuck — physically, by a lugubrious, slow-mo. I don't know if it was brain integration or steady employment, or an intelligent friend, or what; but I had held onto this job — even though I was still slipping into sadness a lot of the time, too.

"I close my eyes — turn the cube to the Right. Open my eyes — turn the Up." And rotate the right and turn and take a step forward into the room.

"And rotate the Up back again and take a step back into your past."

"And slice the cube down and take a step back to look behind you. And slice the cube right to go around an obstacle beside you. Step — slice. Step — slice."

The Cubist experience started getting deeper, more philosophical. It was the power of ambiguity, raised to the exponent of the sign of science; at every turn it was imbuing our lives.

The universe was thinking us into existence, so we could be the intelligent being observing its unfolding and marvel at it. To me it was the tesseract cube within a cube: a point where the three coordinates of space meet the dimension of time — it makes things interesting, just enough complexity so that evolution can unfold. I recalled how Matsen used to say, 'With three parameters you can create an elephant; with 4 you can make his trunk wave.' He was riffing off something Linus Pauling said.

So that we can listen, we evolved in this holonomic, autopoietic, group-plectic, neural-network evolved, algebraic-semantic, bio-semiotic, self-recursive, game-playing automata-stabilized universe to become these quantum-coherent, cross-

temporal, teleo-cybernetic-feedback-entanglement stabilized beings. We are the coordinates in which it is centered, we are carriers, and we are given long enough to feel some of its rays. And we are given long enough to be penetrated by the colors of its rays.

In early models of the world within a world within a world — the spherical harmonics, the music of the spheres — the ancients knew the ratios of the paths of the planets and the sizes. We are in the light cone. Rotation around is moving up and back in a projection into another space.

In Linear Algebra and Hilbert Space, the concept of vectors and dimensions intrigued me. The role of basis is very important in this vector toy world, by a simple change of basis one can change the entire game — domain, universe of consideration. The same vector moving through analogical domains. The analogy featherbeds on a previously known domain.

The coordinates in Levi-Strauss's functional vector space are concatenated in the Canonical Formula:

$$F_x(a) : F_y(b) :: F_x(b) : F_{(a-1)}(y)$$

The myth that you have come to be in gives you a viewpoint. It is your window on the world, partly what you were born with —the family myth, who you are supposed to be. If we consider self [ego, anima, persona, shadow] as a generator of myth, we can see the Levi-Strauss space as isomorphic to Jungian space. The Image of Imager Imaging.

$F_x(a)$: is the anima that is opened and how we are here on the world, drawn to its Other.

$F_y(b)$: is the ego, how we see ourselves conducting ourselves in the world, the presence.

$F_x(b)$ this is like a mytheme of the new myth coming in, the new you that you want to be, being born on the earth.

$F_{(a-1)}(y)$ is the shadow that seeks to reverse, but also is the force that begins the change to overthrow the old order of $F_y(b)$.

We are Cube. Thought is projected on these moving screens by a projector in the nth corner, a projector that is at the intersection of a coalition of dimensions, dimensions connected to senses, and summing, averaging, — reasoning as it were, about inputs.

We are Cube. Inside the Cube, the 3x3 slab of square facets became like windows to look out of, as it is projecting other worlds out of its windows. It projects, and is projected upon. A model for other worlds. So the dimensions are variable, have different basises. This movie house has many projection rooms, but we only buy a ticket to one screening.

I look around, and see — my own eyes: on walls, across floors, within doors — into dreams.

$F_x(a)$ our longing opens up an attention, to something we want to know.

$F_y(b)$ and it is new. Use will make the old obsolescent. And bring the day to light before us, as our mouths open to give us voice and create our cry outside.

$F_{(a-1)}(y)$ or like how things shift back and we are born on the earth. Everything comes by way of these openings. The Canonical Formula is how we bridge the gap of unknowing. It is the inference engine of the savage mind.

We are here and know we are here. And are known through breaks, cuts, chakras, lacunae, slices in a twisting, turning surface. A kind of calculus on a surface but a fractal, osmotic, diffusional musical calculus. If you could stand off from it, abstract from it, you could see it. For now, let us model it by this simple cube within a cube.

Things are born in these openings, be they by eyes, by mouth or the female sex. They are openings, portals, into the world for us, for us to delve into, for us to take the plunge into being. And the idea in living is to open ourselves up to more slices, to be open to the openings of the body, to be more open in our body, to be more open to let more in,

to be attuned to and resonate with. The more we let in the more we embrace the universe. Through our lips, our eyes, our mouth, we are taking part in something that is on the way somewhere, that is being born and dying everywhere at every moment. It is being born like a child inside us, within us and without us. The universe is looking in the interstices, in the spaces between, the bardol spaces, as the soul transmigrates, — looking with an eye for the cunt to be born from, looking with the mouth of the eye to articulate.

We can understand phase space, phase space.

We had the insights of Feynman, so generous and clear, like love from this Peter Pan of chromodynamics, who never lost touch with the element of play.

The surface of the light cone is the space of field quanta that never land. Peter Pan says you get to Neverland by flying 'past the 'second (star) to the right, and straight on till morning' for many days; though the children find their own forms in the landscape of Neverland by believing and looking for them. But really, Neverland is on the surface of the light cone. The real things, objects of this world made out of slower matter are inside the light cone. And outside the light cone are things that are not observable, though possible. But at the boundary between the ordinary world and the superluminal, is the surface — of the light cone. The light cone serves the objects that light and other field quanta can reach in the now instant. It touches them with causality, makes them observable, and relays their attributes to our senses. The entities in the surface of the light cone, photons, gravitons, (they used to be called massless particles, but now they are called connections or strings) move only at the speed of light — never faster, never slower, and they never land.

One day LB and I talked about psychons, kind of like mythemes, theoretical elementary particles of mind. The cubons would be psychons, and they would be arranged in a

lattice of perceptrons. Rather the alignment of perceptrons made a psychon. Perceptrons were more elementary than psychons.

This is your brain. This is your brain on cube: the recursive holonomic fractal universe using self-similar turns of its energy up and down scales and bases.

The dipole of water mediating the phase space. The yellowness of the yellow is bright, all the colors are bright but they don't need to be colors they could be anything, screens upon which are projected scenes, or they could be windows to look through, out /in. From outside, the world is reflected in them.

We are Cube. Our gentle relaxed hands reach out to clasp the cube like a mendicant in a new kind of prayer. The player handles the cube with relaxed dexterity, as he starts to rotate and twist it.

The Cube is a like a society governed by the mechanism of the Invisible Hand, interweaving multiplication of associations. There are like-associated niches, some of which have to be broken up for the good of getting the whole cube going. That is the idea of Buddhism — enlightenment for one leading toward enlightenment for all.

LB and I would be talking.

"Well what would enlightenment look like."

"Oh, I know. It is perfect impedance matching. A person would be in phase with the holistic ocean undulating all around him. He would rise and fall as it rose and fell, he would be buoyed up on it. It would flow through him as though he were in phase. He would be able to transfer the signal from the Great Designer or whatever you want to call it, the Generosity, without impeding it.

I used to feel like that as a kid some time, be on some kind of spiritual high. I felt like I was just in time, able to move in phase with the transparent forces of feelings and

destiny moving through the entities in the outside world. Like the wind, but these were feelings shifted into view points, feeling colors and the other attributes of ontology. I could project myself onto the world. Just as we children were projected by our parents. I don't know how to describe it.

Roots growing in the dark — only growing away from the light — white roots, lacy veils, digging down deep. White clouds carrying the whiteness and quiet of falling snow, white milk in a cat's bowl. Arctic white, beluga white, white wolves and polar bears, terns and hares. Icebergs and bright sunlight. The whiteness of the alb. White clouds in a blue blue sky.

The convexity of the eye lens, round like the world, but made out of parts, like the cube.

Cube, photon, massless particle, field quanta, tachyons of the tychismal, always moving at the speed of light. They are the carriers of information to the eyes, to the touch — about the world to be sensed. Imagined as impossibly small massless round particles, of light, or gravity, — monads to tell us what's what, if we have the phenomenology for it.

Blue jeans, bluebirds, Texas blue bells under the vast and expanding blue sky. Which is due to oxygen in the air getting excited by the presence of the sun, causing its outer orbitals to vibrate in the blue frequency. The true north strong and blue. The ocean is a dark blue; is this due to oxygen too? We are interacting with light photons in the boundary between here and there, then and now; always on the move. I have blue eyes.

Round like the world ball, like the light photon.

At least I think they are round. No one has ever seen a photon — that would be the ultimate uncertainty.

Green as the lawn, as lettuce, as lime. The lovely, lively, life-color. Caterpillars look like the leaves they love.

The hand of man, coming, to shape, to concave, to hollow out, to make an enclosure, to capture what was being carried on the light. The invisible hand that writes, and having writ,

moves on.

Orange as a carrot, as a basketball, a jack-o-lantern with an orange yellow flame. Fire. The orange juice in the morning glass, or a robin's breast — so looked for as the sign of spring. Consequently, Orange is cousin to the sun as seen in the lava that flowed out of the core.

The day deity designer used his hands to shape the eyes, he made them flat like the landscape he could see and touch in front of him. The night deity designer used eyes to model the volume of the space in which he was — but which could not be seen in the darkness.

Yellow is the color of the sun, most beloved by all living things. Consequently, yellow is in the center of our sight because the sun is the center of our existence, and the eye evolved to most easily see the objects with which the generosity has populated our space by staying close to this median frequency of light. The yellow envelope of a bill, the yellow cab to the hospital, the yellow danger sign and the slip on the yellow banana peel. The yellow dump truck, the ugly yellow school bus, the workers and kids in their yellow rain coats. For most visibility, yellow against blue, the most easily understood, unless it is white against blue.

The irises close up at night, and the deity-of-the-night designer has the system go off-logic for a while — just a kind of natural neural chaos, because you need chaos to have attractors.

Red barns and red bricks and fire trucks and fire hydrants. The red lips of the woman in the red dress, the heart that leads to those lips. Red peppers and tomatoes, strawberries and radishes. For red is the kingdom of Christmas most distinguished from the green. All the red stop signs all in a row, blocking your flow. The red ruby and the little red wagon taking you for a ride. Mars is the red planet, symbolizing anger and war, its crust is in flux, it

spews out red dust from its many volcanoes.

The hand that made the irises was a slow invisible hand which reached into everything and it used what it needed from both the day and the night. The designer that made the flower iris also made the eye iris. The petal is a kind of hand for reaching out and touching the color. The iris flower shapes itself into a bowl— to cup, to contain, to imbue, with more time. To organize against entropy. To ratchet the demon of sorting up toward complexity.

We are Cube. Sometimes I like to imagine Cube as a collection of personality traits Some cubons like to associate with other cubons, and they have to be broken up, the comfortable feeling of being next to their natural neighbor has to be sacrificed so that other cubons can associate, so that more can associate, more order can be built up. This is how efficacy comes about.

We are Cube. And the little cubes are like the little Individuals, part of the State, and the (Invisible) Hand comes along and moves them into other relationships. And there, they might get comfortable. But before you know it, the Moving Hand of Fate comes along and moves them again.

Sometimes I like to imagine the little sub-cubes of the array face as rooms I have lived in or been in my life. Lying in bed for example, one could rotate back out of the current room into another room, say one in Quebec, or in San Antonio.

Think of all the bodies in all the rooms of the world: this still wouldn't come close to the number of configuration possible with the Cube. What if somehow in dreams during the night we could freely move back and forth in each other's spaces.

I would frequent a shack on the shore, white faded porch, light curtains wafting on a breeze through the window. Perfect for dreaming.

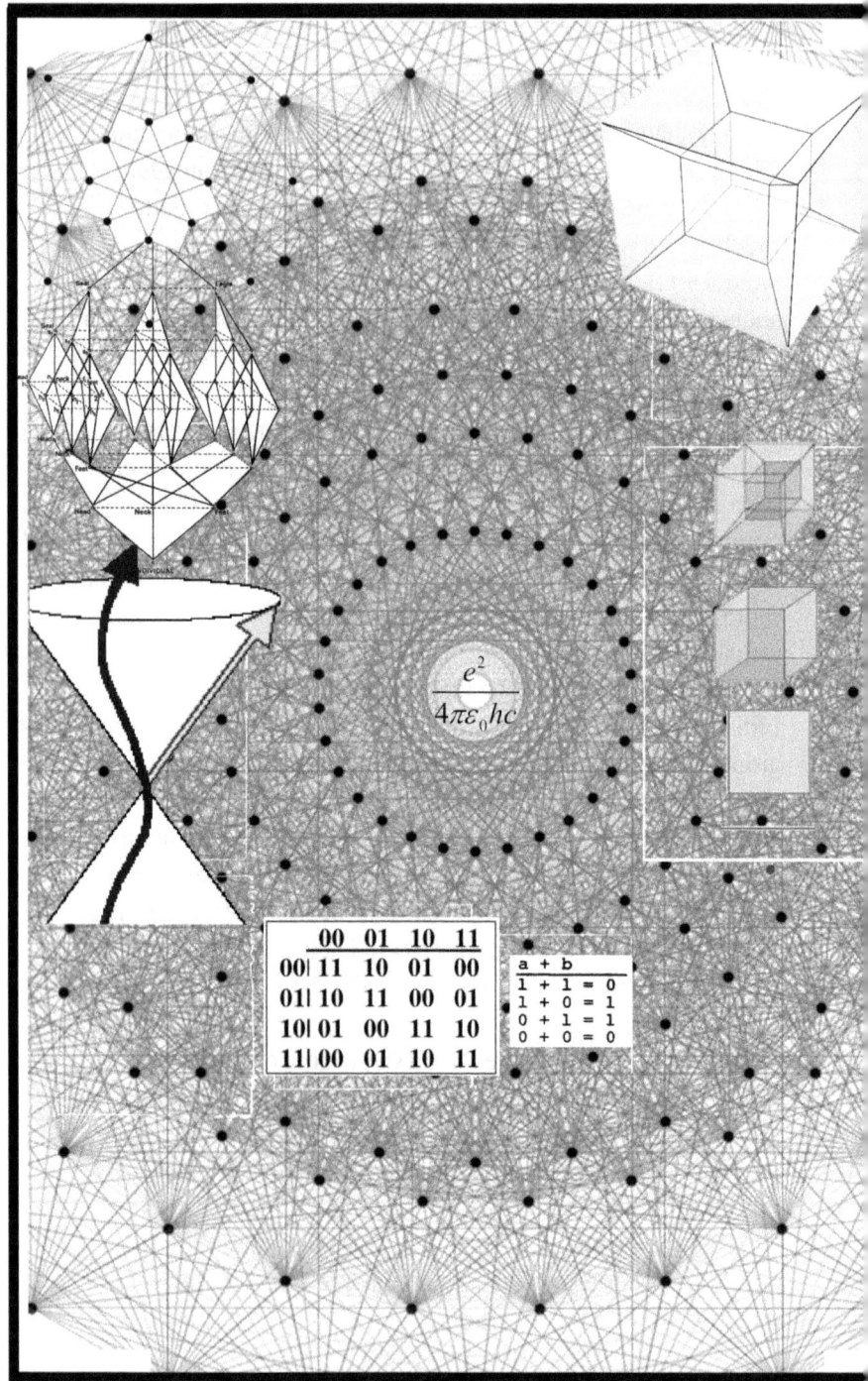

The Cascading Projections of Universal Symmetry through Groups into M

And we are way downstream in the light

We know time, from light, and
from an averaging of several times.
The observers evolved on a planet
dominated by a 24 hour circadian rhythm.
And a larger seasonal one,
and a much smaller one in the heart.
Not to mention all the exquisitely intricate timing circuits,
— some that make their own memory storage,
for a more relaxed exploration of time.
Time is not just an endless march of seconds
going in lock step from t_0 to the end.
It is part of the object, its existence.
Though it is sobering, that rocks and aluminum
are around a lot longer than we are,
we know time.
We convolve time; we are time binders.
Through us, the now is brought to light.

From the light up stream we come,
on a path, through a hole in the night.
And we are way downstream in the light.

We come from the depths of dark space
and have evolved these eyes to see back into the night
and forward into the day.
Though we can not see ourselves
here in this between-ground.
We emerge on the surface
of a manifold generosity dispersed by the sun.

The general strategy is to grow a surface around the entity:
bark around the tree, skin around the hand,
this beautiful pellucid vitreous humor around the eye . . .
Time is always local, occurring at a focal point
in some confined region.
Like different points of view,
in the poem, the narrative, the theatre:
of flesh and of the soul,
through which "we" pass —instantiations
of an unfinished form.
We reconstitute the light.
The light brings the now.
Coming out of street lamps, and flash lights,
turning corners in head lights,
glowing — undulating in swimming pools,
reflecting off jewels and passing through glass,
and the night.
Time is a cone of light,
like the way a lamp projects a cone
of light up onto the ceiling
and down onto the floor.
The dark outside the cone is the possible but not yet,
for it can not be reached by the light of now;
that inside the cone, touchable by the light, is the real.

There exist many time zones,
enfolding any object like a bubble.
The object starts to have its own time:
it encompasses every object
as it begins to exist;
around any swelling in a surface,
around any opening in a fold,
around any situation in passing.

And all these times are in touch with each other
the way the river is in touch with itself
— up stream and down stream and all along its banks.
It is in touch with the rocks and trees that impact its flow,
and force it to feed back on itself
in little swirling eddies
which are minute convolutions of itself
in the serpentine flow of the river of deep time.

From the light up stream we come,
on a path, through a hole in the night.
And we are way downstream in the light.

And we are here because of the sex of the female.
They tested some young women in math
and found that if they gave some of the girls
a picture of a baby,
they did not score as high on the math test.
Which is not to go all *post hoc,*
ergo propter hoc on them and say:
mothering instinct occludes abstraction.
Correlation is not necessarily causation.
If I took anything away from statistics, it was that.
Anyway, who cares?
Compared to being the eye
through which the generosity sees
to the occurrence of intelligent observers in the world,
being good at math is a pale comparison.

The now expands out — a pregnant bubble
around everything big and small,
enclosing the universe itself —
through which we fall —

slowly in a line of least action, plowing a field,
going to work, sitting in lawn chairs.
Walking by the sea. (We come from the sea.)
And every one of us
recapitulates our phylogeny in our ontogeny.
We see time in action,
like the wind blowing clothes on a line.
We are frames of reference,
pages on a spine, turning in the horizon.
Cleaving space with our presence
as our wishes become guesses . . .
It is there when we go to sleep inside it.
The night ponders and we dream.

From the light up stream we come,
on a path, through a hole in the night.
And we are way downstream in the light.

We are all time binding agents
in a vast, intricate plot,
an immense journey through the dark into the light.
We have eyes that evolved
and are expressed on our face
to embrace the great leap we have made across space.
We now know time to be like light fragmented,
through the interstices of symmetry of stained glass windows
and thence to be reified in the mind's eye,
of those who are watching it intensely.

From the light up stream we come,
on a path, through a hole in the night.
And we are way downstream in the light.

My Years of Apprenticeship at Love

Sex is the Antigravity of Metamorphosis   1

Set in Montreal, Vermont, and Austin, 1975-76 the story follows Walker Underwood as he tries to settle down with a young mom in the city of his origin. Experience falling in love, the responsibilities of being a father figure, the encounter with the sign system of Other — that is love's apprenticeship.

ISBN 0-965584291	2008
ISBN 978-0965584296	$25

The Indigenous Tribesmen of Neverland   2

Set in Austin 1979, follows Walker Underwood in his relationship with high school girl Laura. Others are Wild Bill, and teen Weesa, Laura's friends. The April-July romance has punks and hippies influencing each other. The insight that group theory brought through physics and anthropology into zeitgeist settles in.

ISBN 0-965584275	2010
ISBN 9780965584272	$25

Dolores Park   3

Set in Berkeley, San Francisco and Mount Shasta, 1981-82. Follows Walker as he pursues romance in a tantric Buddhist sex commune. There the exegesis of living in sangha employ love and envy to assault and grind on the ego as romantic precepts fall. Blueprint for running a new age commune.

ISBN 0-965584232	2001
ISBN 978-096558423	$25

a sextet of novels by Michael Lyons

Seeing through the Spell of Transference

Set in Berkeley follows Walker Underwood, his pasionate love affair with girlfriend Ruth and psychoanalysis of his mind with his therapist Zenobia Kafka. Lota sex in this book. Works with the idea of talking as cure for the malais of soul. Talking book movie with programming outline.

ISBN 0-965584240   coming 2011
ISBN 978-0965584241 $25

A Blue Moon in August

Set in Berkeley and San Francisco and Mexico, 1990-91 story follows Walker as he goes decides to marry and gets to know his new wife in Lamaze class. Follows Walker as he become a new father under spell of motherhood. Explores cyberpunk writing and the possibilities of parental romance.
ISBN 0-965584259   2005
ISBN 978-0965584258 $25

Thoughts on Vacation

Set in San Francisco, and the California Sierras, 1991-2000 the story follows Walker as he become a new father. It is a comedy of manners about raising children and how we are raised by them. Mr. I is a streamlined information space will raise your IQ if you let it. Explores mindset of a Silicon Valley information worker.
ISBN 0-965584267   2005
ISBN 978-0965584265 $25

"Little House on the Prairie" Trilogy

Cultivating the Texas Twister Hybrid

Set in Austin, 1978 follows Walker Underwood and Greg and their 3 dogs in great detail about setting up and operating a pot farm. Like a trip back into the 1800s. Bust at the end.

ISBN 0-965584208   1998
ISBN 978-0-9655842-0-3 $20

The Secret of the Cicada's Song

Agreatly expanded chapter of the first book in the series Cultivating the Texas Twister Hybrid. This book is a peyote trip in prose and poetry and a strange moving symbolic language based on the distinction operator of G. Spencer Brown. First to use C syntax poems

ISBN 0-965584216   1998
ISBN 978-0-9655842-1-0 $20

Knight of a 100 eyes

Agreatly expanded chapter of the first book in the series, *Cultivating the Texas Twister Hybrid*. This book is the three time frames of a Tai Chi long form. The parallel times of the 30 minute set, the growth of begining to journeyman player, and the life time of witness to the great Tao, as Ecology, as Gravity, as Form of Fractals. It is a modern commentary on the I Ching.
ISBN 0-965584224   2002
ISBN 978-0-9655842-2-7 $25

Ebook Availability
We are currently exploring eBook options.
Check Hitmotel.com for Available Titles
Ebooks cost 1/4 of the paperback list and are available on the Web, iTunes and other places.

Ebook devices
- Computer,
- iPhone, iTouch, iPad
- Android, Sony Reader
- Kindle
- Other Mobile Devices —

http://www.hitmotel.com

A CD of recorded poems, radio plays, readings is avilable at Hitmotel. Also available on the Youtube channel

http://www.youtube.com/hitmotel

An unknown and transparent figure of international letters, Michael Lyons once lived in Austin. He has authored a dozen literary books. This novel is the 2nd volume of the "My Years of Apprenticeship at Love" sextet. It is the 5th of that series to be published by Hitmotel.

http://www.hitmotel.com **HiT MoteL Press**

The "Little House on the Prairie" trilogy

The "My Years of Apprenticeship at Love" sextet

Poetry at HiT MoteL Press

I learn to walk again
Slow baby steps after a serious skateboard accident.
Sequel to How I spent my Christmas Break

Happy Trails to the Infinite
Fourteen Sonnets
The influence of form in every-day life.

http://www.hitmotel.com

Diamond Head
Return to the place of the Honeymoon on a family vacation. Learn surfing, and the secretes of the sea.
Do to things what light does to them.

Chap books and collection available **FREE** in .pdf at www.hitmotel.com

Selected Poems
This is a selection of some 80 poems going back to the 70s up to the present. The poems are selected from all the books of Michael Lyons, including some rare chapbooks.

The poetry is usually the personal observations of the self in the world with others. There are some sonnets.

The poems are loose and spontaneous and usually humorous.

The Punctual Actual Weekly was a literary magazine in Berkeley of 1976. This memoir of the artists and writers in the community it served presents reprints and reviews. In addition to the solo theatre presentations, the magazine format supports a graphical novel treatment of a poet's notebook, and an art show catalog of works. Besides the Abstract Expressionism, Physical Theatre, and (humorously) applying the insights of several isms: structuralism, imagism, surrealism, actualism, personalism, set in the city of Berkeley (a philosopher known for idealism) it attempts what WCW did in Paterson.

ISBN 0-965584283   360pages   2007
ISBN 978-0-9655842-8-9 $30

HiT MoteL Press
www.hitmotel.com
These books can be ordered from any book seller or on-line.
They are deeply discounted on Amazon, and Barnes& Nobles.
Check www.hitmotel.com for selections, recordings, to order.

Boho Novels, Memoirs
The "Little House on the Prairie" Trilogy:
Cultivating the Texas Twister Hybrid, a portrait of the artist as a weed gardener (1998) ISBN 0-9655842-0-8 $20.00
The Secret of the Cicadas' Song, a peyote trip in poetry and prose (1998) ISBN 0-9655842-1-6 $20.00
Knight of a 1000 eyes, about Tai Chi, movement, Laban, and the I Ching (2002) ISBN 0-9655842-2-4 $25.00
others:
The Punctual Actual Weekly, about the life and times of a small mimeograph literary rag centered around artists living in a Berkeley warehouse and the Amphictionic Theatre (2007) ISBN 0-9655842-8-3
The Church of the Coincidental Metaphor, youthful adventures in Mexican radio

Novels: The "My Years of Apprenticeship at Love" Sextet:
Sex is the Antigravity of Metamorphosis, tales of romance and hitchhiking in North America. (2008) ISBN 0-9655842-9-1 $25.00
The Indigenous Tribesmen of Neverland Bohemian life in Austin slacker enclaves. (2010) ISBN 0-9655842-7-5 $25.00
Dolores Park, Texan joins a California Tantric Buddhist commune (2001) ISBN 0-9655842-3-2 480 pages. $25.00
Seeing throught the Spell of Transference A cab driver's journal of psychotherapy. ISBN 0-9655842-4-0
A Blue Moon in August, about marriage and children late in life. (2005) ISBN 0-9655842-5-9
Thoughts on Vacation, a father is raised by his child and is enlightened by mortality. (2005) ISBN 0-9655842-6-7

Check into HiT MoteL @www.hitmotel.com for cover art, interactive Table of Contents, e-book sample chapters, recordings and other mindware.

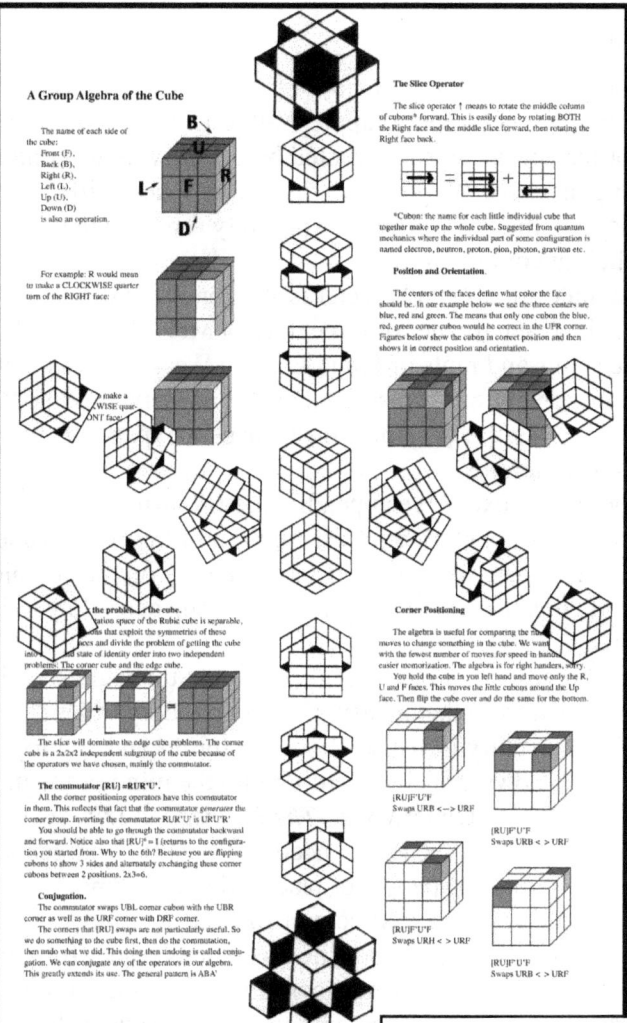

A Group Algebra of the Cube

The name of each side of the cube:
Front (F),
Back (B),
Right (R),
Left (L),
Up (U),
Down (D)
is also an operation.

For example: R would mean to make a CLOCKWISE quarter turn of the RIGHT face.

make a CLOCKWISE quarter ONT face

the problem of the cube.
ation space of the Rubic cube is separable, os that exploit the symmetries of these state of identity enter into two independent problems. The corner cube and the edge cube.

The slice will dominate the edge cube problems. The corner cube is a 2x2x2 independent subgroup of the cube because of the operators we have chosen, mainly the commutator.

The commutator [RU] =RUR'U'.

All the corner positioning operators have this commutator in them. This reflects the fact that the commutator generates the corner group. inverting the commutator RUR'U' is URU'R'
You should be able to go through the commutator backward and forward. Notice also that [RU]⁶ = I (returns to the configuration you started from. Why to the 6th? Because you are flipping cubons to show 3 sides and alternately exchanging these corner cubons between 2 positions. 2x3=6.

Conjugation.

The commutator swaps UBL corner cubon with the UBR corner as well as the URF corner with DRF corner.
The corners that [RU] swaps are not particularly useful. So we do something to the cube first, then do the commutation, then undo what we did. This doing then undoing is called conjugation. We can conjugate any of the operators in our algebra. This greatly extends its use. The general pattern is ABA'

The Slice Operator

The slice operator † means to rotate the middle columns of cubons* forward. This is easily done by rotating BOTH the Right face and the middle slice forward, then rotating the Right face back.

*Cubon: the name for each little individual cube that together make up the whole cube. Suggested from quantum mechanics where the individual part of some configuration is named electron, neutron, proton, pion, photon, graviton etc.

Position and Orientation.

The centers of the faces define what color the face should be. In our example below we see the three centers are blue, red and green. The means that only one cubon the blue, red, green corner cubons would be correct in the UPR corner. Figures below show the cubon in correct position and then shows it in correct position and orientation.

Corner Positioning

The algebra is useful for comparing the s moves to change something in the cube. We with the fewest number of moves for speed in hand easier memorization. The algebra is for right handers, s
You hold the cube in you left hand and move only the R, U and F faces. This moves the little cubons around the Up face. Then flip the cube over and do the same for the bottom.

[RUF]F'U'F'
Swaps URB <--> URF

[RUF]F'U'F'
Swaps URB < > URF

[RUF]F'U'F'
Swaps URH < > URF

[RUF]F'U'F'
Swaps URB < > URF

A Group Algebra of the Cube,
a short cube solution is available free.

http://www.hitmotel.com

Books vailable wherever fine books are sold
Or on-line,
Or ask at your local library,
Order a personalized copy from the author at:
http://www.hitmotel.com

The Indigenous Tribesmen of Neverland

A novel by Michael Lyons

The novel celebrates the joys of Austin slacker living. It continues the narrative of hippie Walker Underwood, in the 30th year of his age as he is confronted by maturity in the context of being in a love relationship with a high school girl, Laura. We move in the lush spring of April-July romance in a low rent Tortilla Flat community.

Another important character is Walker's friend Wild Bill, a messianic avatar, owner of three dogs, a Melville figure. There are the songs and rants of Laura's friend Weesa, a punk Lolita. The transgenerational cross fertilization of inspiration goes both ways. Together they explore consciousness expanding epiphanies of this generation.

There is an extensive appendix titled A Structural Analysis, which has commentary on Castaneda and Levi-Strauss and a solo theatre piece with aesthetic commentary on the amphictionic. It ends in a post-modern text growing according to its own rules.

The biggest story of my generation was the understanding that unfolded from the diffusion of group theory in the later half of the 20th century into disciplines of physics and anthropology. This novel shows that story settling into the mind of the main character, Walker Underwood. *The Indigenous Tribesmen of Neverland* explores the two concepts of the title: Indigenous tribesmen — natural people of place, ad hoc community of peers, reflecting kinship, structuralism; and Neverland — the Peter Pan Syndrome, of romantic bohemians who refuse to "grow up" and become part of consumer corporate culture. Neverland is also associated with the field quanta of the surface of the light cone, that carry information about the now to senses.

The Indigenous Tribesmen of Neverland is the 9th novel by Lyons. It is Volume 2 of the "My Years of Apprenticeship at Love" sextet. It is the 5th book of that series to be published by Hitmotel Press.

Literature / Romance / Humor / Metaphysical / Theatre

340 pages   ISBN10: 0965584275 $25
                 ISBN13: 978-096558427

www.ingramcontent.com/pod-product-compliance
Lightning Source LLC
Chambersburg PA
CBHW060109170426
43198CB00010B/822